POULTRY AILMENTS
FOR
THE FANCI

GW00630643

More Advanced Work on Poultry Diseases

For those who wish to study poultry diseases in greater detail reference should be made to:

Poultry Diseases Under Modern Managememt,
G S Coutts, Veterinary Surgeon

Available from the publishers.

POULTRY AILMENTS FOR THE FANCIER

JOSEPH BATTY (Editor)

Beech Publishing House
Spur Publication Books Ltd
7 Station Yard
Elsted Marsh
Midhurst
West Sussex GU29 0JT

© Joseph Batty, 1997

This book is copyright and may not be copied, stored or
reproduced in any way without the express permission of
the publishers in writing.

.ISBN 1-85736-316-7

First Published in this form 1997

Beech Publishing House
Spur Publication Books Ltd
7 Station Yard
Elsted Marsh
Midhurst
West Sussex GU29 0JT

CONTENTS

FOREWORD

After many requests for a concise guide on poultry ailments the present short work has been compiled. Although not an exhaustive or definitive book on the subject, it does attempt to cover the main ailments found by the poultry fancier who keeps birds. The normal fowl, large and bantams, are included, as well as many of the aspects of showing. Ducks, geese, turkeys, peafowl, pheasants and other Game birds are also part of the contents.

The approach is a combination of first hand experience in dealing with ailments in my own birds and the experience of veterinary surgeons who have written on the subject. The first part is based largely on a section which appears in *Practical Poultry Keeping*, edited by myself, but originally written by the staff and advisors of Poultry World. Where necessary, new notes have been added on developments on the common diseases such as Marek's which is causing considerable concern to fanciers who keep highly inbred breeds such as Sebrights and Silkies. The work of a prominent expert in poultry diseases has also been used.

My experience has been that Coccidiosis is the worst problem and therefore great care must be exercised to provide sheltered accommodation, avoiding wet or damp conditions. Often this means keeping chicks indoors until the danger is past. Turkeys are quite vulnerable and should be reared off the ground or in hygienic conditions inside. Clean shavings and adequate water and food are essential; moreover, the food given should contain the drugs needed to combat Coccidiosis or Blackhead (Turkeys). The correct degree of warmth for rearing is also essential because if too hot or cold the chicks will be more likely to pick up an infection.

Sound management will avoid most of the problems and when birds do occasionally suffer some problem this should be spotted quickly and dealt with by medication, moving to fresh ground when the old ground becomes sour. In the case of a serious outbreak the affected birds should be isolated and antibiotics or other treatment obtained. If the birds are valuable then a Veterinarian should be consulted and antibiotics obtained. Much of the work can be done by the average fancier once the problem is recognized.

Only keep bird if they can be given the correct attention which will be a relatively short time each day. Constant vigilance to detail on food and water and the avoidance of overcrowding will ensure that disease is kept to a minimum.

Joseph Batty

Plymouth Rock
(Barred)

Polands
(Gold)

Redcaps

Rhode Island Red

Scots Greys

Sussex
(Speckled)

Charles Francis

SOUND MANAGEMENT
VITAL
FOR
DISEASE CONTROL

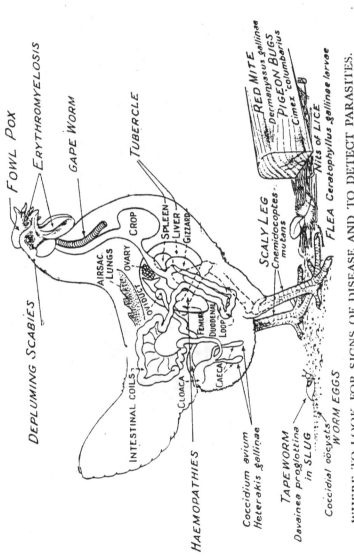

WHERE TO LOOK FOR SIGNS OF DISEASE AND TO DETECT PARASITES.

1

SOUND MANAGEMENT VITAL

Basically there are two main systems of management:

1. **Intensive** (Battery cages or intensive on deep litter)
2. **Free Range** (Birds allowed outside to fresh air and grass)

There are also a variety of systems which fall in between the two extremes; giving some of the features of free range, but in a limited way.

LARGE SCALE CAN INCREASE DANGERS

Intensive forms of management such as **straw yards,** and built-up or **deep litter** may have much to recommend them as labour saving systems, but, nevertheless, they increase the risk of the spread of infectious diseases and parasites as a result of the denser populations, which increase the opportunity for infection to be spread, especially where communal and automatic feeding and watering devices are in use. Respiratory diseases in particular spread more rapidly under fully intensive conditions where there is little dilution of droplet or airborne infection by outdoor air.

In recent years there has been general acknowledgement that birds kept in cages suffer from brittle bones and many other ailments which impose suffering on the stock. Attempts have been made to increase cage sizes and add refinements such as perches, thus allowing birds to move around better and go to roost as in natural conditions.

The greatly publicized **free range system** is not free from diseases and ailments, but at least birds running around can eat natural food and enjoy fresh air, ass well as exercise which is vital to sound health. Birds running out also build up resistance and, provided they have adequate shelter in inclement weather can achieve high results. There is the danger of disease from wild birds and from ground which is not disease free and this aspect must be watched very carefully or any outbreak will spread rapidly.

Fanciers who keep a few birds can see each individual and often pick up birds at least once a week so their condition can be felt – a plump, yet not fat – bird should be the aim.

Disease control, therefore, cannot be achieved without a sound knowledge of husbandry and hygiene.

Essential requirements are :

1. **The correct spacing of birds,**
2. **Adequate ventilation,**
3. **Prompt recognition and isolation of ailing birds,**
4. **Proper management of litter and**
5. **General sanitation of housing and equipment.**

These are just as important, if not more so, than knowledge of modern drugs.

It must be remembered, too, that subnormal health resulting from chronic infections, parasitism and, more commonly, **faulty nutrition**, interfering with full production and efficiency, can take an even greater toll of profits than the more spectacular and acute killing diseases.

Economists have also shown that **mortality ranks second to food costs** in the expense of producing eggs, and can exert a more serious effect on the profitability of table poultry production than an increase in food costs.

Until recently, advice in the case of most outbreaks of infectious disease has been to destroy infected or ailing birds, depopulate and disinfect. This advice, still sound in many instances, was based on the fact that the low value of the individual fowl was such that the health of the flock as a whole was of more importance than that of any of its component units, and treatment of the individual was uneconomic and dangerous.

The introduction of vaccines and specific drugs that can be used mixed in the food or water has made the prevention and treatment of a

Healthy Producer

Pin point Pupils

Pearly Eye

Eye Showing Tape-worm

Head & Eye Signs -- best to stick to normal eyed birds.

number of diseases a safe and practical proposition. It has meant that the accurate diagnosis of the nature of the disease is even more important than hitherto, for these new drugs, highly powerful in their effect, are equally highly selective in their action against different disease agents and are not "cure alls".

A drug highly efficient against one disease may be equally valueless against another. Its faulty use, therefore, resulting from wrong diagnosis may be not only wasteful but also dangerous in that the disease may go unchecked. It would be useless, for example, to attempt to cure a condition resulting from a vitamin deficiency by the use of a drug for coccidiosis, or to hope that a vaccine for fowl pox would prevent coryza. Yet such beliefs exist, and not only cause untold damage, but sooner or later may bring the drug or vaccine concerned into disrepute.

Recognition of Symptoms

One cannot labour the point too strongly that **the first cardinal law in hygiene or disease control is the rapid recognition of the sick bird and the establishment of an accurate diagnosis.** Fortunately, it is usually possible in poultry keeping to sacrifice a typically affected bird for post-mortem examination. Although the poultry keeper should acquaint himself with the lesions or changes which accompany the commoner diseases so that temporary measures can be put in force, he should as far as practicable always obtain a qualified diagnosis, for many diseases can be differentiated only by laboratory techniques.

Probably the second most important point in hygiene is that birds which have been newly purchased or returned from other premises and which may be in an incubative stage of a disease should be isolated for at least three weeks before being mixed with the home flock. Purchases of new stock, of course, should always be made through reliable and reputable sources, and preferably not through auction markets, dealers or pet stores. Many outbreaks of fowl pest are traced to the latter sources. For commercial purposes chicks should be purchased only from flocks where routine blood testing is carried out.

The live bird is, of course, the greatest danger in transmitting disease, and in addition to the foregoing precautions it must be remembered that adult fowls often harbour parasites without themselves showing symptoms, and that they can transmit such infection to the more susceptible chicks. Chicks should be reared, therefore, as far as is practicable, in isolation from adult stock and separate rearing ground and equipment should be kept for this purpose. Traffic of attendants between the old and young stock should be reduced to the minimum. These precautions appear to be particularly valuable in the control of fowl paralysis.

Carriers of Diseases

In a similar way certain species of poultry can carry diseases which affect others. The **caecal worm** of the fowl, for example, is responsible for spreading the parasite that causes blackhead in turkeys, and for this reason it is undesirable to attempt to rear turkeys either in contact with fowls or on ground where poultry have recently been kept.

The carcasses of fowls dying from an infective disease are also common sources of infection. They should be removed immediately from the runs and never left lying about to be attacked by vermin, which could in turn transmit infection to the rest of the flock. **For preference such carcasses should be incinerated.** Lime pits are sometimes used, but are not so efficient. *Never leave carcasses lying on the manure heap.*

Discourage any neighbours or other farmers from bringing sick birds or the carcasses of dead birds on to your premises for examination, and avoid visiting premises in order to advise on outbreaks of disease. That is the province of the veterinary surgeon, and in any case there is the danger that you may transmit infection back to your own stock.

Rats and mice, apart from being carriers of certain diseases such as Salmonellae infections, can do a lot of damage and cause serious waste. **Food stores should, as far as possible, be vermin proof.**

Clean egg production is essential in the control of egg borne disease, since bacteria in dirt and droppings can spread through the pores and infect the embryo, which in turn will affect other chicks during hatching. This is quite apart from the lower returns from packing stations for dirty eggs or the labour involved in cleaning them. This is an important point, for no egg by whatever method it is cleaned is ever as clean bacteriologically as an egg which is laid clean. Many methods of cleaning eggs increase the risk of spoilage during storage, since the washing of an egg in any fluid of a lower temperature than that of the egg causes bacteria to be drawn through the pores of the shell. Wiping with a damp cloth always hastens penetration of bacteria, while the cloth itself becomes contaminated and spreads infection to other eggs. A danger with washing machines is that the brushes may become contaminated and transmit infection to subsequent eggs.

Dry cleaning, although laborious, is probably preferable to wet cleaning, or, alternatively, dirty eggs can be dipped in germicidal solutions containing a detergent. It is essential to remember, however, that the solution so used must be at a temperature higher than that of the egg – i.e. 80 to 90 deg F. Clean egg production can be improved by collecting eggs at more frequent intervals from the nest boxes, and seeing that the nest box litter is renewed frequently.

Spread of Diseases

Most infectious diseases are spread by food or water which becomes contaminated with infected droppings and other discharges. With the exception of the built-up or deep litter system, therefore, droppings should always be treated as being potentially infected. **This means either the use of droppings boards, which should be regularly cleaned, or the removal of droppings from contact with the bird by wire floors, droppings pits, movable folds, etc.**

Some parasites – coccidia, for example – are not infective immediately they are passed out by the host, but must spend a period outside the bird's body before they can infect a new host or re-infect the same bird. In the case of coccidia, this period is, in the most favourable conditions for the parasite, a minimum of 48 hours. Advantage can be taken of this fact by removing folds every other day and cleaning droppings boards etc., at the same interval.

Warmth and humidity favour the multiplication of such parasites and damp areas in the litter should be avoided either by moving water and feeding troughs frequently or using one of the sanitary types of feed and water troughs which prevent birds having access to their droppings.

Grass in runs should be cut short so that parasites and other infective agents get a maximum exposure to sun and light, rapid drying being one of the surest methods of destroying infection. Coccidia, on the other hand, can resist lower temperatures, remaining in a quiescent state, but will again multiply and become infective when climatic conditions are favourable. This is the reason why outbreaks recur on farms from season to season. Worm eggs are also highly resistant, and remain infective for long periods in the litter or soil.

Sterilising the Soil

Disinfection of the soil is difficult and uncertain, and the best method of dealing with infected ground is to plough or dig over the land, lime heavily and re-seed, leaving it vacant for the longest possible period. Fixed houses, therefore, should have alternative runs which can be regularly rested. Remember, too, that bushes, ditches and the like can harbour slugs and insects which may act as the intermediate hosts for parasites such as tape worms. Some of these sources can be destroyed by dressing the runs with copper sulphate.

Dealing with Parasites

External parasites such as lice and mites appear to multiply most quickly in a dark humid atmosphere. Houses, therefore, should be well ventilated and well lit, and they should be so constructed that no part can be the permanent harbour for dust and dirt. Perches, nest boxes and droppings boards are the most favoured sites for external parasites, and these should be movable so that the fittings can be dismantled for regular cleaning.

In the same way, walls and floors should, as far as is practicable, be smooth and free from cracks so that they can be easily washed. Treating the walls, floor and ceiling with creosote, lime washing or paraffin emulsions incorporating a strong insecticide is an effective method of controlling external parasites.

Disinfect Twice a Year

A thorough disinfection of all poultry houses should be carried out at least twice each year, and always before new stock are moved into the house. It must be remembered that most disinfectants lose their efficiency in the presence of organic matter such as grease and manure, while bacteria and viruses are so minute that they can be

protected in the smallest piece of dried excreta. The first stage, therefore, in disinfecting a house is to sweep up all litter and refuse, scrape
the floor, walls and ceiling and remove the accumulated sweepings to
the manure pit. Where there has been infection present in the house,
then it is advisable to spray first with a strong disinfectant then, having
collected the litter and sweepings, to burn them.

The empty house should then be scrubbed with hot water containing 4 per cent washing soda or a detergent. A high pressure pump
is the most suitable for this purpose, since it will penetrate to inacces-
sible places. After drying, the house should be sprayed with an ap-
proved disinfectant in a strength recommended by the manufacturers.

Dealing with Litter

Built-up litter, of course, is a different proposition and it depends entirely on the use to which this is being put, as to the steps that
should be adopted. In the case of adult birds, the litter can be used for
a number of seasons provided that disease has not been present and the
litter has been well managed. It is inadvisable, however, to attempt to
rear young stock on built-up litter that has previously maintained
older birds.

In the case of chick rearing for broilers, it is still debatable
whether the litter should be removed and renewed between each batch
or whether it should be heaped in piles approximately 1 metre high and
1.75m. at the base for a period of about a week before being re-
spread. Heaping of built-up litter does produce temperatures capable
of destroying parasites such as coccidial oocysts and worm eggs. Even
when litter has been heaped, however, it is advisable to spread fresh
litter under the hovers. Regular turning of the litter helps to turn fresh
droppings to the base where they will be exposed to quantities of
ammonia lethal to coccidia.

Generally speaking, so far as the control of parasites is concerned, the deeper the litter the better, right from the start. Shallow
litter increases the risk of setting up outbreaks of disease.

Risks from Overcrowding

Obviously, in considering the control of disease, the density of
population must be taken into account, since overcrowding will increase the potential spread of infection, while inadequate hopper space

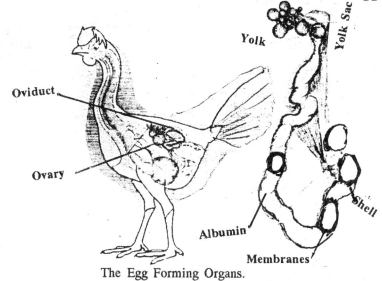

The Egg Forming Organs.

Adapted from *Poultry Diseases* B F Kaupp

or perch space can have a similar effect.

The disinfection and hygiene of incubators and hatchery premises is a subject on its own and outside the scope of this chapter. In general terms, however, the method most commonly in use is by generating formaldehyde gas by mixing 40 per cent formalin with potassium permanganate in a wide-mouthed jar or bowl. The quantities recommended for routine use are 3 oz formalin to 2 oz potassium permanganate for every 100 cu ft of incubator space. This is somewhat above the amounts previously recommended in the Poultry Stock Improvement Plan, but it has been shown that these quantities are required for the certain destruction of bacteria. If disease has been present 5 oz formalin and 3 oz potassium permanganate should be used.

Incubation Hygiene

Eggs should **not** be fumigated during the 24th to 84th hours after setting. In routine hygiene it is usual to fumigate immediately the eggs are set and again after the hatch has been taken off. Where separate hatching compartments or hatchers are in use the ports should be closed during fumigation. Fumigation of the hatcher should be

Anatomy of the Hen

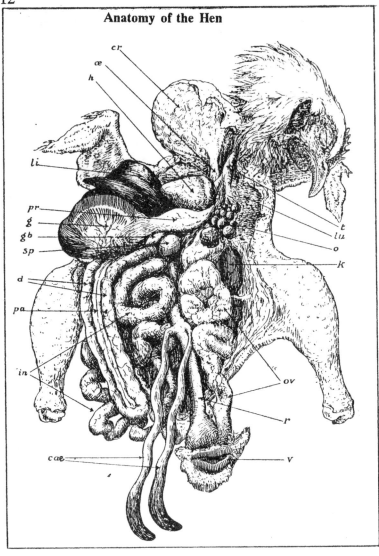

cr—crop ; œ—gullet or œsophagus ; h—heart ; t—windpipe or trachea ; lu—lungs ; li—liver ; o—ovary ; k—kidney ; pr—glandular stomach or proventriculus ; g—gizzard ; gb—gall-bladder ; sp—spleen ; d—duodenal loop (bowel) ; pa—pancreatic gland ; in—coils of gut or intestine ; cæ—cæcal pouches ; ov—oviduct ; r—rectum or last part of digestive tract ; v vent or cloaca.

Bruff Jackson.

carried out before removing the fluff and debris, and such debris should be burnt.

Removable parts of the incubators, including the trays, should be washed with hot washing soda or other detergent or hypochlorite solution.

Particular attention should always be paid to any second-hand equipment. Dealers' crates should not be permitted on the premises, while food should be obtained in sterilised sacks or in non-returnable paper bags. Chick boxes should be non-returnable.

ENSURING PRODUCTS ARE FRESH

The entire poultry industry is concerned with food and ensuring that this is made available to the consumer quite fresh and without danger of transmitting disease to humans or other users. Unfortunately , eggs and other produce deteriorate quite quickly and within 7 to 10 days they may be unfit for human consumption.In other words, they are perishable products and must be treated as such.

Steps suggested for eggs are :

1. Store eggs in a special storage room at below 20 ° C (the temperature of hen) and preferably around 15 ° C , thus ensuring that they keep fresh as long as possible and this should start immediately.

2. The store should be equipped with fans to remove stale air and cool and there should be no windows to let in sunlight. A refrigerated room is too costly and not desirable.

3. Wash eggs only when dirty and then in water which has not been used too long and in a detergent which will not contaminate the eggs.

4. Eggs should be moved for delivery in specially cooled vans.

5. Date stamp eggs with the date of laying or the date of recommended consumption.

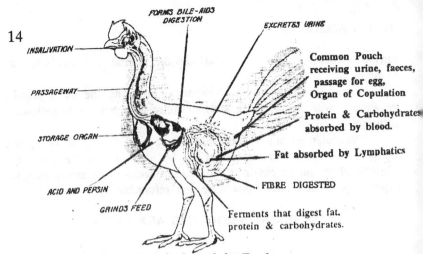

INSALIVATION

PASSAGEWAY

STORAGE ORGAN

ACID AND PEPSIN

GRINDS FEED

FORMS BILE-AIDS DIGESTION

EXCRETES URINE

Common Pouch receiving urine, faeces, passage for egg, Organ of Copulation

Protein & Carbohydrates absorbed by blood.

Fat absorbed by Lymphatics

FIBRE DIGESTED

Ferments that digest fat, protein & carbohydrates.

Functioning of the Fowl
An understanding of the food food conversion is vital

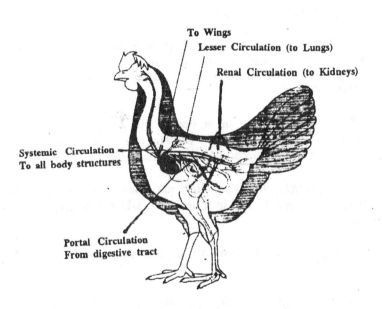

To Wings

Lesser Circulation (to Lungs)

Renal Circulation (to Kidneys)

Systemic Circulation To all body structures

Portal Circulation From digestive tract

Circulation of Blood in a Fowl

6. A quality guarantee mark (eg, little lion should be stamped on the egg). This is a controversial issue because the small free range producer may not have access to an official stamp -- yet his eggs may be the freshest of all.

Meat processing also requires strict control throughout the entire production process and regulations regarding food processing which are extremely strict and with appropriate official inspections.There should also be regular checks made to ensure that all flocks, whether laying or table birds are not suffering from salmonella or other diseases.

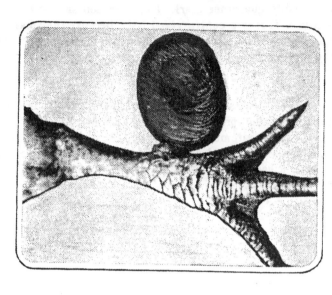

Overgrown spurs may do a lot of damage to the breeding hens.
harmless, one with a soldering iron and the other with
Here are two tips for making the spurs

Faverolles
(Salmon)

Hamburgh
(Gold Pencilled)

Houdan

Indian Game
(Dark)

Leghorn
(Cuckoo)

Nankin Bantams

2

DISEASES AND PARASITES

For convenience and clarity, the diseases of poultry can be divided into two main groups. These are:

1. Specific diseases; i.e. those caused by specific agents such as bacteria, viruses and other parasites and which are infectious or contagious - and

2. Non-specific diseases which include nutritional deficiencies, constitutional disturbances in individual birds, and conditions associated with breeding, management or environment.

SPECIFIC DISEASES

This is probably the most important group since, in view of their infectious nature, they may spread rapidly through a flock and cause severe losses. The diseases in this group can be sub-divided into **bacterial, viral, protozoal and parasitic.**

BACTERIAL DISEASES

PULLORUM DISEASE

This is more commonly known as bacillary white diarrhoea or B.W.D., and is caused by a germ, *Salmonella pullorum.* It is probably one of the most serious causes of losses in baby chicks, the mortality varying from 20 to 80 per cent.

The symptoms are not constantly characteristic and in acute outbreaks may be absent, chicks being found dead in large numbers under the hovers. In some cases death is preceded by lack of appetite, listlessness, continual chirping and coma; a white diarrhoea is not constantly present.

Post-mortem findings are equally variable and in acute cases there may be little to be seen. In some instances small white nodules appear in the lungs, in the liver and on the heart wall. In all cases, however, specimens should be sent to a veterinary laboratory for confirmation by a bacteriological examination.

A high proportion of chicks which survive an outbreak become carriers, and although they may appear to be healthy the majority of such cases will harbour the germs in the ovary and other organs. When these carriers mature, a proportion of their eggs will contain the germ and infect the embryo.

During incubation the germs multiply in the embryo and embryonic fluids, so that when the chicks hatch out and as they dry off, bacteria are carried round in the air currents of the incubator by infected down and other excretions and debris. This in turn may infect large numbers of chicks hatching at the same time.

Chicks remain highly susceptible up to about three weeks of age and infection can be transmitted from diseased to healthy chicks by droppings and other excretions contaminating food, water and litter in the brooder house, or in chick boxes, etc. Infection can also be spread by the sexer's hands, second-hand utensils and so on.

Treatment for this disease is possible is the drug furazolidone, been used at the rate of 0.04 per cent in the food for 10 days, to give a high percentage of recoveries, and the majority of chicks so treated will not be carriers of the disease. Treated chicks should be isolated and fattened for table or kept for commercial egg production, but not used for breeding.

There are two methods of controlling this disease. The first depends upon elimination of carriers from the breeding stock and this is carried out by means of a blood test, which detects such carriers. There are two methods of undertaking the test:

(a) The rapid whole blood test, which is carried out in the field, using a stained antigen. When a drop of blood from an adult infected bird is mixed on a porcelain plate with a measured quantity of antigen, antibodies in the carrier's blood will cause the bacteria in the antigen to agglutinate together into well marked clumps. In the case of a non-infected bird the blood-antigen mixture remains unchanged.

(b) The serum tube test, in which blood is taken from the bird and sent in a tube to a laboratory. At the laboratory serum is removed from the blood and mixed in a tube with a suspension of bacteria. This mixture is then incubated and, again, a carrier is denoted by a clumping or agglutination of the bacteria at the bottom of the tube.

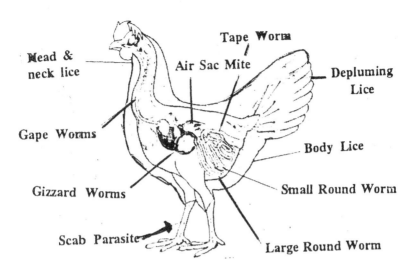

Head & neck lice
Tape Worm
Air Sac Mite
Depluming Lice
Gape Worms
Body Lice
Gizzard Worms
Small Round Worm
Scab Parasite
Large Round Worm

Common Parasites and their location

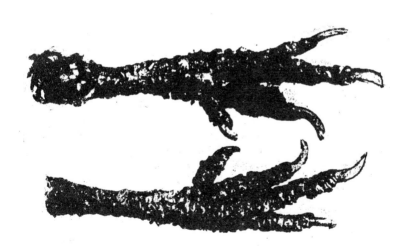

Scaly Legs of two years standing .
Caused by a parasite and occurs on free range birds.

In healthy, non-infected birds the mixture remains unchanged.

Details of testing can be obtained from any veterinary labora-tory, veterinary surgeon or County Poultry Advisory Officer.

After removal of the carriers, disinfection of the houses should be carried out and the remaining non-reactors re-tested at monthly intervals until no further reactors are found. Furazolidone has been found to be of value in the treatment of such chronic carriers.

The second method of controlling the disease depends mainly on incubator hygiene, which has already been discussed in the preceding chapter.

SALMONELLOSIS

This term is used to denote outbreaks of disease in chicks, turkey poults, ducklings, etc., caused by organisms of the Salmonella group other than *S. pullorum* and *S. gallinarum* (fowl typhoid). It is one of the largest groups in the bacterial kingdom in this country. Most species, however, are comparatively rare, and the majority of outbreaks are caused by *S. typhimurium, S. enteriditis* and *S. thompson.*

In many respects the disease is similar to Pullorum disease. It occurs in young chicks up to three weeks of age, the mortality again being variable and the symptoms and post-mortem findings are not characteristic. Diagnosis can only be made on a bacteriological investigation. Survivors of an outbreak also become carriers and harbour the organism either in the reproductive tract or, more commonly, in the intestinal tract.

A portion of the eggs laid by such carriers will again be infected, and in many instances infection arises from contamination of the outside of the shell with infected droppings. As the egg cools, particularly if cleaning is carried out by washing in fluids at a lower temperature than the egg, then bacteria are sucked through the pores of the shell and set up infection in this way. From then on the disease spreads in the incubator and brooders as in Pullorum disease.

Unfortunately, this group of organisms occur very commonly in nature and cause disease in a variety of species of animals including man. They are particularly common in rats and mice and many out-

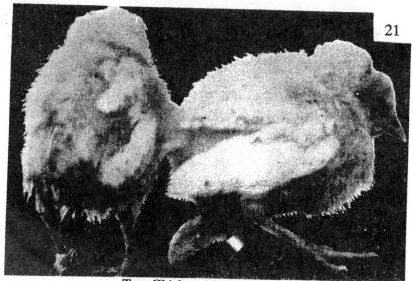

Two Chicks with BWD

Bird infected with Fowl Pest

breaks probably originate from feeding stuffs, etc., contaminated by infected vermin.

Blood testing is not entirely satisfactory in detecting carriers in breeding flocks so that control of the disease must mainly depend on incubator and brooder house hygiene.

Clean egg production is one of the most effective methods of controlling the disease, together with fumigation and the use of germicidal egg dips.

FOWL TYPHOID

This disease is also caused by an organism of the Salmonella group known as *S. gallinarum*. It differs from the previous two diseases, however, in that in the majority of instances adult birds are affected and outbreaks in young stock are comparatively rare. It may be acute or chronic, and the mortality rate may vary from 20 to 80 per cent.

Infected birds show lack of appetite, profuse yellow diarrhoea, paleness of the head and usually die within forty–eight hours of first appearing ill. In more chronic cases symptoms may last for a few days. On post–mortem examination the liver is congested, bronze-green in colour and the lungs dirty brown. In chronic cases nodules are found in the heart and the wall of the intestines.

Treatment of the disease can be effectively carried out with **furazolidone** at the same dosage as already mentioned, but the success of the treatment depends on removing the treated birds to clean land and also on the speed with which treatment is started.

Carriers of the disease can be detected by the same blood test as that described for Pullorum disease. Control of the disease, however, depends on improved hygiene. The carcasses of dead birds should never be left lying about, but should be burnt or buried in quicklime. Houses and equipment should be thoroughly disinfected. Vaccines are available, but are of doubtful efficacy. **Furazolidone** at a level of 0.04 per cent in the ration is the most effective drug so far known for the treatment of this disease.

TUBERCULOSIS

In poultry this is caused by the avian strain of the tubercle bacillus. The strain is also responsible for tuberculosis in pigs, and for this reason pigs and poultry should *not* be kept in close contact. **Infection occurs by the contamination of food and water with infected excreta.**

As in other animals the disease is usually chronic in nature and infected birds become emaciated, pale and anaemic and may show a persistent diarrhoea. In some instances lameness may be present. On post-mortem examination typical tubercles, i.e. white, cheesy nodules, are found in the liver, spleen and intestines and in the joints.

No treatment is available and infected birds should be destroyed. In valuable stock an attempt may be made to eradicate the disease by the use of the tuberculin test, which is carried out by injecting tuberculin into the wattle. In infected birds the wattle becomes hot and swollen within 24 to 48 hours. After testing, the non-reactors must be moved to disinfected houses on clean land and re-testing carried out at intervals.

In most cases, however, the most economic procedure is to dispose of the infected flock. With improvements in hygiene and management the disease has become less common in recent years and is most commonly seen in second and third year birds.

FOWL CHOLERA

This is a highly infectious, fatal disease and may cause mortality up to 80 or 90 per cent. It is caused by the bacterium known as *Pasteurella multicida.* Normally the disease is absent from this country and the majority of serious outbreaks have been due to imported stock from Europe.

The disease is rapid in its course and in many outbreaks few symptoms are seen, apart from congestion of the head parts, dullness and profuse green diarrhoea. Post-mortem examination may be inconclusive, and bacteriological investigation is required for confirmation.

Treatment is not recommended in this country, and in most outbreaks it is advisable to destroy the infected birds and to carry out thorough disinfection. A chronic type of the disease occasionally occurs, when the wattles become infected and appear swollen and wrinkled. Mortality with this chronic or wattle form is low, though occasionally individual birds may die suddenly at irregular intervals..

The chronic form is also known to occur in turkeys, when treatment can be successfully carried out with injections of antibiotics. Turkeys, geese and ducks are all highly susceptible to the acute type.

INFECTIOUS ARTHRITIS

Outbreaks of this condition are occasionally seen in young growing stock, causing severe swelling of the hock joints and lameness. The disease is due to infection of injuries in the food or leg region with *staphylococci*. The original injuries may be due to factors in the management such as glass, wire, stubble, thistles, etc. Mortality is seldom high.

Treatment can be effectively carried out by injections of antibiotics, together with removal of the birds to a new environment.

INFECTIOUS CORYZA (Contagious Catarrh)

A mild form of this disease is caused by the bacterium *Haemophilus gallinarum*. Although losses are usually few, the majority of birds in the flock may become infected and there is usually severe economic loss as a result of interference with growth and egg production.

Infected birds usually show a thin, watery discharge from the nostrils and eyes; the incubation period is short and the disease may spread through the flock in a matter of days. A number of environmental conditions, particularly faulty ventilation, over-crowding and malnutrition (particularly a deficiency of vitamin A) predispose towards infection.

Treatment can be carried out with sulpha drugs, particularly sulphathiazole and sulphamezathine. The addition of disinfectants, such as hypochlorites and Lugol's Iodine, to the drinking water, helps to cut down cross infection. Attention should be paid to the various predisposing factors mentioned.

CHRONIC RESPIRATORY DISEASE

The exact cause of this disease is thought to be a pleuropneumonia-like organism. The disease which is similar to coryza, is more serious and more chronic in its effect. The incubation period is

Fowl Cholera
Note Swollen Earlobe

Fowl Roup

prolonged, and outbreaks may exist in a flock for some weeks or even months, and deaths may occur, in addition to a more marked effect on production and growth.

Affected birds in the early stages show nasal and ocular discharge. This discharge tends to dry up into firm, cheesy masses which block the nostrils and sinuses of the head, causing distortion and swelling of the face and often blindness. In advanced cases the infection spreads to the deeper parts of the respiratory tract, and it is thought that the same predisposing factors as those mentioned in the case of coryza play an important part in the spread of the disease.

Treatment can be carried out in the early stages by injections with antibiotics, and there is some claim that the feeding of antibiotics at high levels is of some value.

Broncho Pneumonia
Distressed breathing with eyes closed and beak wide open.

3

VIRAL DISEASES

FOWL POX

This disease is sometimes known as chicken pox, avian diphtheria or diphtheritic roup. It is contagious and occurs in two forms. The first, known as the **skin type**, is shown by the occurrence of small, watery blisters on the comb and wattles, round the corners of the beak and more variably in the thinly feathered parts of the skin. These blisters tend to dry up into brown crusts and may run together to form large cauliflower–like growths.

The second form of the disease, the **diphtheritic type**, is shown by the occurrence in the mouth and the tongue of dirty-yellow, cheesy membranes, which when removed leave a raw bleeding surface.

Infection is usually spread by contamination of small wounds in the head and mouth region. Red mites may spread the disease, and it is more commonly seen in unsanitary conditions of housing. Although the death rate may not be high, large numbers of birds may become infected and have to be destroyed.

Provided the disease is observed before it has spread through the flock the most satisfactory control is to destroy the affected birds, remove the non–infected birds to clean houses and clean ground, and to vaccinate. It is inadvisable to attempt treatment unless the most rigorous isolation of the treated birds can be effected.

Vaccination is carried out by brushing a drop of vaccine on to the lightly scarified skin of a small area of the thigh from which the feathers have been removed. Immunity is not produced until about fourteen days after vaccination and lasts for about four to six months.

INFECTIOUS LARYNGO TRACHEITIS

This is a highly infectious disease for which the mortality rate can be exceptionally high. The disease usually occurs suddenly, the bird showing symptoms of coughing, sneezing and extending its neck and making a prolonged inspiration through the open beak. Breathing is accompanied by rattling, and blood clots may be coughed up and

seen on the walls of the house. On post-mortem examination, blood
and casts of clotted blood or cheesy material will be found obstructing
the wind pipe. A chronic form of the disease in which the mortality is
low also occurs.

There is no known treatment of value and vaccination is not
recommended in this country. Infected birds should be destroyed and
disinfection carried out. Most outbreaks probably originate from contact of healthy birds with carriers, since a recovered bird may harbour
the virus for over year.

FOWL PEST

This is a collective term used, for legislative purposes, to include two diseases, **Newcastle disease and fowl plague**. The latter is
not known to occur in Great Britain. Fowl pest is compulsorily notifiable and if confirmed the infected flock is usually slaughtered (depends on Regulations in country concerned) , compensation being paid
for the birds not infected at the time.

A number of supplementary orders to the *Fowl Pest Order* have
been made from time to time in an attempt to eradicate the disease.
These Orders include restrictions of movement in various districts,
closure of markets, Orders dealing with the importation of hatching
eggs and livestock, the establishment of clean and black areas, and so
on.

This disease occurs in two main forms. The acute type is highly
infectious with a mortality rate often as high as 90 to 100 per cent. In
this form death usually occurs in two to three days of the first appearance of symptoms of a frothy, yellow diarrhoea, purplish congestion of
the head and comb and a high pitched, rattling cough.

In the sub-acute form there is usually a sharp and severe drop in
egg production with a tendency to the laying of misshapen or shell-
less eggs, which are frequently laid on the floor of the house instead of
in the nest box. There may be symptoms of colds, nervous twitching
of the head and neck and bending of the head back over the body or
towards the ground.

**If any of the above symptoms are suspected the outbreak
must be reported immediately to the police.** While diagnosis is
pending other poultry keepers should not be allowed on the premises
nor to come into contact with the poultry.

The ailing birds, or even apparently healthy birds from a suspect flock, must not be sold or sent to market. If the disease is confirmed slaughter or other mandatory action is carried out, and the necessary disinfection, etc., is supervised by the veterinary staff of the Ministry of Agriculture.

The acute form of the disease usually results from birds having access to infected material such as swill, hotel waste, etc., and all material likely to contain poultry scraps must be boiled for at least one hour before being used. The mild form of the disease is commonly spread through markets and dealers.

AVIAN LEUCOSIS COMPLEX   **(Including Marek's Disease).**
Earlier books referred to this term to describe a group of diseases, including fowl paralysis, visceral and ocular lymphomatosis and various forms of leucosis, and which were wrongly thought to be different manifestations of the same disease.

Recent work, however, has shown that there are three, if not four, quite separate disease entities caused by different causal agents and probably not associated one with another. The use of a variety of pseudonyms such as lymphomatosis, leucosis, leukaemia, big liver disease, fowl paralysis and the like, has given rise to a great deal of confusion, even within the literature; and despite the vast amount of research work which has been carried out on the leucosis complex, our knowledge regarding the exact cause and methods of transmission of these conditions is still incomplete.

The main disease coming under this heading is **Marek's Disease** which is a form of paralysis occurring in wings and legs. Even exhibition poultry such as Silkies and Sebright bantams are affected and the cause is the invasion by lymphocytes (white blood cells) and the development of tumours.

Fortunately some breeds and strains are immune so they do not develop the problem and, if there has never been any problem, this may continue. In other cases the chicks must be vaccinated without delay (at a day old) using an *Herpes vaccine*. Otherwise the disease will develop and remain dormant until the point of lay when the stress brings on the condition. Hens are more prone to the problem than males.

Marek's disease is now regarded as the most dangerous of the viral diseases and many of the earlier conditions are covered by that disease. The remainder are diseases which appear similar, but are not and the Herpes vaccine does not prevent these other diseases. On one point, however, there is complete agreement, and that is that the complex is undoubtedly the commonest single cause of loss in poultry and accounts for nearly forty per cent of the deaths in mature birds in this country. Possible types are:

(1) AVIAN LEUCOSIS . There are three forms of this disease recognised, depending on the type of blood cell involved, but its division into three forms is somewhat academic. All three forms are neoplastic – that is, they are cancerous and consist of the abnormal multiplication and subsequent infiltration and deposition into various organs of different types of immature blood cells.

(a) **Lymphoid leucosis.** This is the commonest form of the disease and is the one usually referred to as visceral lymphomatosis. It occurs in two forms, either discreet or diffuse. In the discreet form there are multiple, white, soft tumour growths, mainly occurring in the liver, spleen, kidneys, heart and lungs. The liver is most commonly affected and may be enlarged to three or four times its normal size.

In the diffuse form, known as *big liver disease*, the liver and spleen are greatly enlarged but retain their normal outline. The cells infiltrate through the whole organ changing it to a mottled, greyish-red or brownish colour.

(b) **Myeloil leucosis** . Again, this may occur either as a discreet or diffuse form. In the diffuse form, again, the liver and spleen are usually affected, generally enlarged, firm, granular and with a marbled, brown appearance.

In the discreet type soft, chalky tumours occur on the inner surface of the breast bone, along the ribs and sometimes in the abdominal cavity.

(c) **Erythroleucosis.** In this form, mature red cells rather than white cells are involved. The liver and spleen are swollen, cherry-red in colour and soft and fragile. There are no tumour masses.

Both types (b) and (c) are easily transmissible and associated with a filter passing virus. Whether the same virus is responsible in both is as yet unknown. Type (a) would appear to be transmissible, but the presence of a virus has not been definitely proved.

Marek's Disease
Should vaccinate at 1 day old

Examples of Liver Affected by Diseases
Revealed by post-mortem examination.

In all three types, egg transmission probably occurs but most workers agree that transmission by the egg is probably, under natural conditions, of less importance than the spread of the disease during early brooding, and the exact part which egg transmission plays is still open to some doubt.

Symptoms of leucosis are very variable, depending on the organ and type of the disease, but it generally occurs in birds from about six months to a year, although cases may occur at a later date. There is loss of condition, usually paleness of the head and interference with egg production.

Although the course of the disease may be chronic, death may occur suddenly. It has been suggested that blood sucking insects may play a part in the spread of erythroleucosis.

No treatments of vaccines are known to be of any value for this disease. It would appear that apparently normal birds can act as carriers and transmit the infective agent in their excreta, or possibly via the egg. Unfortunately no accurate test is known by which such carriers can be detected and removed from the flock.

(2)FOWL PARALYSIS. This disease, also known as neurolymphomatosis, is now considered to be a chronic, inflammatory, infectious condition and unassociated with leucosis. The disease produces enlargement of various nerves, particularly those to the wings and legs, to the intestinal tract and the chest wall.

In some instances these nerve lesions may be accompanied by "tumour–like" growths in various organs, particularly the ovary and less commonly the liver, spleen, kidneys, etc. These tumours may be confused with the discreet form of lymphoid leucosis.

Symptoms will depend on the nerve involved, but the commonest is a paralysis of the legs which starts with a limp followed by a clutching appearance of the food and finally complete paralysis of one or both legs, when the bird may die as a result of inability to reach food. Dropping wings, twisted neck, impaction of the intestines and difficulty in breathing may occur when the nerves associated with these organs are involved. Fowl paralysis occurs at an earlier age than leucosis, i.e. three to six months.

(3) IRITIS OR OCULAR LYMPHOMATOSIS. The latest information suggests that this disease is quite separate from the two already described and that, therefore, the term ocular lymphomatosis no longer

applies. The condition is possibly infectious but the actual cause is obscure. The iris of the eye loses its normal colour and fades to become bluish-grey, while the pupil becomes irregular and slit-like, and loses its ability to respond to light.

(4) OSTEOPETROSIS OR MARBLE BONE   This disease, which is relatively uncommon, was also previously thought to fall within the leucosis complex but it is now considered to be, again, a separate entity. The cause, like that of iritis, is obscure but it appears to arise from interference with normal bone metabolism and may be associated with a number of factors. The disease consists of gross enlargement, distortion and thickening of the long bones, mainly those of the shank.

Although the exact mode of transmission of leucosis and paralysis is obscure, there is fairly strong evidence that chicks are most susceptible during the early rearing period and that the greatest spread of the disease occurs during the first two or three weeks of life. As a result it has been shown that the greater the degree of isolation in which chicks can be reared from adult stock the lower should be the incidence of the disease.

The question of egg transmission as far as fowl paralysis is concerned has not yet been proved. It is thought that certain factors such as coccidiosis, may predispose towards infection with fowl paralysis.

It is known that families vary in their resistance to the leucosis complex and that such genetical differences among families can be sued to reduce losses by either mass selection or by progeny testing.

The question of breeding for resistance, however, is extremely involved, and the cost is such that it is doubted whether many breeders could attempt it. Moreover, there is the added disadvantage that it involves retaining diseased birds on the premises to which the progeny can be exposed in order to test their resistance.

There is no known treatment for the leucosis complex., except prevention by vaccination for Marek's disease. Birds which still contract paralysis should be killed and incinerated. If they have been vaccinated against Marek's disease then it would indicate that one of the other diseases is involved.

PSITTACOSIS

This is a virus disease mainly occurring in parrots and related species of birds. A form of the disease, however, has been identified in pigeons, ducks and chickens. In view of the fact that the parrot form can cause fatal disease in man, the disease is dealt with by legislative Order involving compulsory slaughter.

DUCK VIRUS HEPATITIS

This is a disease which occurs in baby ducklings causing a mortality rate as high as 90 per cent. The disease is sudden in its onset and has a rapid course of symptoms lasting little more than an hour.

The infected ducklings appear sleepy, have their eyes closed and death follows in a few minutes, occasionally preceded by falling over on their sides, kicking spasmodically. Ducklings over four weeks of age do not appear to be infected.

On post-mortem examination the liver is enlarged and covered with haemorrhages, the blood vessels to the kidney are congested and the spleen also enlarged.

No effective treatment is known , but an antiserum can be used to prevent the disease.

EPIDEMIC TREMOR

This disease occurs in chicks between one to four weeks of age and occasionally in slightly older birds. There is an unsteadiness which develops into paralysis of the legs. The chick becomes inactive or sits on its hocks, frequently falling over on its side. Tremor of the head and neck is observed and general debility increases followed by death in a few days.

The condition is sometimes confused with so-called "crazy chick disease". There is no known treatment or prevention, but the feeding of a special diet containing Vitamin E can prevent its occurence.

Leg Paralysis

Limber Neck

UNCLASSIFIED DISEASES

PULLET DISEASE OR BLUE COMB

This is a common disease in this country although the exact cause is obscure. It is thought that a low grade virus is responsible and that this virus may be present on most premises, but that disease becomes apparent only when some other predisposing factor such as a sudden change of food or environment acts as a "trigger" mechanism. Recent work suggests that infection may be spread in the droppings.

Although the disease occurs acutely, mortality is not usually high, although the majority of the flock may become infected at the same time. There is sudden loss of appetite accompanied by a white diarrhoea and the head parts are a dark colour.

The most striking symptom is a sudden drop in egg production which in pullet flocks may decrease from 50 per cent to nil in a matter of two or three days. Affected birds show a tendency to depraved appetite and will eat coarse grass or litter leading to crop impaction. Though a few birds may die, the remainder recover spontaneously in three to four weeks.

The disease results from breakdown of the kidneys with deposits of urates occurring in the internal organs. It has been found that mild laxatives such as Epsom or Glauber salts and the addition of 10 per cent molasses to the mash are of value.

The addition of 0.5 per cent potassium chloride to the drinking water has also been recommended, and more recently high level feeding with antibiotics has been claimed to be effective.

It must be remembered that a drop in egg production also occurs in fowl pest and when this is experienced the outbreak should always be reported in the first place.

4

PROTOZOAN DISEASES

COCCIDIOSIS

This disease is **the commonest cause of loss in chick rearing,** being caused by a minute single-celled parasite called the *coccidium*. These parasites attack the cells lining the inner wall of the intestinal tract and produce disease in the acute forms by rupture of the blood vessels of the affected bird, while in the less acute cases there is inflammation of the lining of the intestines.

There are many different kinds of coccidia occurring in a wide variety of animals and each species produces its own type of disease. Coccidia are host specific, i.e. each variety of animal is only attacked by its own particular type of cocciodia which in turn is incapable of affecting another type of host. For example, coccidia of the fowl are quite distinct from those affecting turkeys. In the fowl a number of species of the parasite occur producing broadly two forms of the disease.

The **caecal, or acute, type** occurs in chickens mainly during the first few weeks of life and can cause losses often as high as 50 to 60 per cent. In this form death is due to haemorrhage by rupture of the blood vessels in the caecal tubes and the affected chicks may pass blood or blood-stained droppings. Chicks stand about listless, with their eyes closed, drooping wings and ruffled feathers. On examination the caecal tubes will be found filled with blood or blood clots.

In the second form, known as **intestinal coccidiosis,** the disease is somewhat more chronic and occurs in birds of from three to six months of age and occasionally older. Affected birds are usually thin and emaciated and may show a persistent diarrhoea, and in older birds egg production is depressed. Symptoms can be easily confused with other diseases, and should be confirmed at a laboratory.

The life cycle of the coccidial parasite is a complex one, but the important point to remember is that the parasite cannot infect a new host immediately it is passed out in the droppings since it must pass a certain period on the ground outside the hosts. This period depends on climatic conditions, of moisture and humidity, but cannot be less than forty-eight hours.

On the other hand, the parasite may remain alive outside the body of the host for as long as twelve months and still be infective when the conditions are again favourable for its multiplication. The optimum condition for the multiplication of the parasite is a temperature of 80 to 90 deg F, *and a moisture saturated atmosphere.* At lower temperatures, even down to freezing, multiplication is inhibited, but the parasite will still remain alive in a non-infective stage. Excessive dryness will also stop multiplication.

Most outbreaks occur from April to July and are particularly severe following mild, wet Springs when multiplication is proceeding at maximum speed. Use can be made of these facts in controlling the disease by sanitary measures such as cleaning out pens every two or three days, using dry, easily swept up litter, or moving outdoor brooders three times weekly.

Food and water troughs should be moved frequently or raised on wire floors so that the chicks' droppings pass out of reach. Attention should also be given to floor ventilation, to the dryness of the litter and to the avoidance of damp areas. Overcrowding should be avoided, for it results in a high concentration of infection being built up.

It is known that the regular intake of a small number of coccidia stimulates the development of natural immunity, and although it is desirable to permit access to a sufficient number of the parasites to produce this immunity it is difficult in practice to regulate the amount of infection ingested. It explains, however why in epidemics some chicks may succumb while a number survive, for the latter will have acquired resistance.

It must be remembered, however, that this immunity is specific to the type of coccidia with which the chick was infected, and a chick which has survived an outbreak of the caecal type and is immune, will not be immune to the intestinal forms. Certain types of management – if hygiene is of a very high order such as wire floored brooders, for example – may produce highly susceptible chicks. Mixing of chicks of different ages is inadvisable, since older birds, themselves immune, may act as carriers.

Chick Affected by Coccidiosis

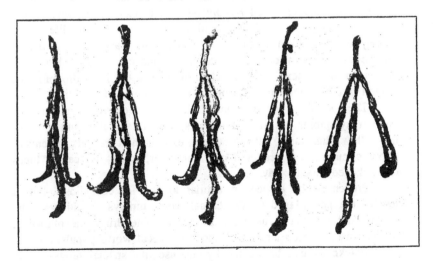

Coccidiosis

The coccidial parasite in the resting stage is extremely resistant to most disinfectants and is most effectively destroyed by a 10 per cent watery solution of ammonia preceded by a thorough scrubbing of the brooder or pen with hot water containing 4 per cent washing soda or other detergent.

For many years now effective drugs have been available for the treatment of this disease. Sulphamezathine at 0.2 per cent in the drinking water was one of the first to be commonly used, and other drugs which are available are sulphaquinoxaline in the food or water at 0.04 per cent, and nitrofurazone at 0.022 per cent. These drugs are available in premix form or in solution.

Care must be taken that the drugs are used at the correct levels and for the correct periods of time recommended by the manufacturers, since excessive doses for prolonged periods of the sulphonamides may give rise to toxicity.

In the prevention of the disease sulphaquinoxaline can be used continuously at 0.0125 per cent, and nitrofurazone at 0.011 per cent. These drugs do not interfere with the production of natural immunity.

BLACKHEAD

This is one of the commonest causes of loss in turkeys and is caused by a parasite called *Histomonas*. The exact life cycle of the parasite is uncertain, but it apparently spends a part of its life inside the egg of the chicken caecal worm and in this way is protected rom climatic conditions when outside the hosts, and can remain alive from one season to the next. During actual outbreaks, the fresh droppings of infected birds are probably one of the main sources of infection.

The mortality rate in young turkeys is frequently high and the common symptoms are ruffled feathers, loss of appetite and a mus-tard-yellow diarrhoea. Post-mortem examination reveals greenish-yellow circular areas scattered over the liver, while the caecal tubes are thickened and ulcerated.

Prevention by hygiene is similar to that for coccidiosis, but in view of the part played by the caecal worm turkeys should never be run on land previously used for chickens. The drug, phenothiazine, has been used in an attempt to reduce the caecal worm population.

Blackhead can be treated by the use of   special drugs-- one called Enheptin at the rate of 0.1 per cent, or in the prevention of the

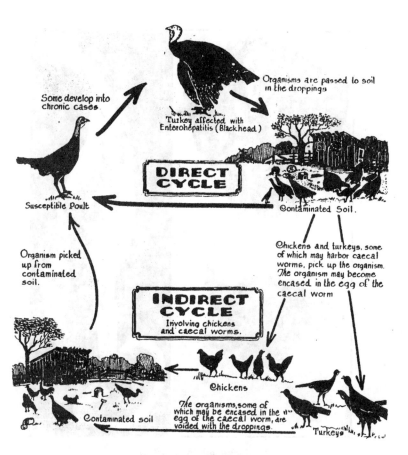

Life Cycle of Blackhead in Turkeys
(after A R Winter)

disease at 0.05 per cent. This drug is available in premix form for adding to the mash. Affected birds can also be individually treated by the drug in capsule form and the drug in solution is now available for medicating drinking water.

Blackhead in Turkeys
Top Left: Infected Liver. *Bottom Left:* Drooping wings & Sunken Eyes.
Right: Right Caecum containing lesions; Left normal.

5

PARASITIC DISEASES

INTERNAL PARASITES (see page 19).

Most of the larger parasites which are found in poultry are only harmful if they occur in excessive numbers or if the bird is already debilitated by other disease. Damp and unhygienic houses encourage multiplication of parasitic worms and the condition appears to be increasing in intensivism such as straw yards and built-up litter.

Large Round Worm

This worm (Ascaris) is about 1.50 in. long and greyish-white in colour and is found in the small intestine. Heavy infections are responsible for poor growth and interference with egg production. Carbon tetrachloride at the rate of 2 cc per bird or preparations containing 1.3 per cent nicotine and 3 per cent phenothiazine are also available for medicating mashes.

Small Caecal Worm

Although common in poultry this worm (Heterakis) is mainly important since it appears to be a vehicle for the blackhead parasite. The worm can be eliminated by phenothiazine or phenothiazine and nicotine.

Gizzard Worm

This worm (Amidostomum) occurs in goslings, causing ulceration of the lining of the gizzard and is frequently fatal. The worm is only about 0.125 in long and hair-like and can be seen only with great difficulty. Treatment can be carried out with carbon tetrachloride in 2 cc capsules.

Gape Worm

The gape worm (Syngamus) is now rarely found in chicks, although outbreaks are occasionally seen in turkeys and in game birds. The worm, which is red in colour and "Y" shaped, adheres to the wall of the windpipe, causing gasping, coughing and death by obstruction. Treatment can be carried out with a powdered drug called barium antimonyl tartrate which is blown into a container into which the infected chicks are placed.

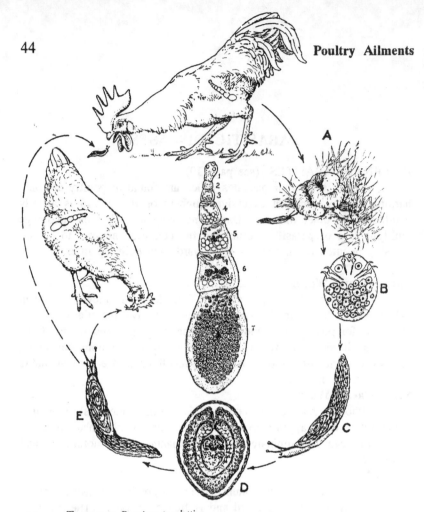

1-7.	Tapeworm *Davainea proglottina*.
1.	Scolex or head with suckers and hooks.
2-3.	Undifferentiated segments.
4-5-6.	Segments with male and female generative organs.
7.	Mature segments or proglottids with eggs; these drop off at intervals and carry on the infestation.
A.	Droppings with proglottids which crawl out and are being eaten by slugs.
B.	Small embryo (oneosphere) armed with six hooks.
C.	Slug containing oneosphere which develops into—
D.	encysted shape which forms in the slug and is eventually eaten by a fowl and develops into a fully grown tapeworm *Davainea proglottina*.

Life Cycle of Tapeworm

Tape Worms

There are two types of tape worm in poultry. The small tape worm (Davainia) occurs in the small intestine and severe infestations cause considerable interference with growth and production. This small worm is just visible to the naked eye and usually can only be definitely diagnosed by a post–mortem examination. The large tape worm is comparatively rare.

Treatment of tape worms is unsatisfactory. The drugs recommended are Kamala 1 gm, carbon tetrachloride 1 to 21 cc, or arecoline. The intermediate hosts of tape worms are slugs, snails and certain insects, and the treatment of the pens with copper sulphate as a dressing or spray is of benefit.

EXTERNAL PARASITES

In the fowl external parasites fall into much the same category as the worms in that slight infestations are fairly common, while heavy infestations can cause severe interference with growth, production, and, in young stock, even death.

The red mite (Dermanyssus) is probably the most serious in its effect, and may cause death from anaemia. It fees on the bird at night. The northern fowl mite (Lyponyssus) is also fairly common. It is blackish in colour and can be seen crawling in the feathers, whereas the red mite when gorged with blood is a bright red colour. During the day the mites congregate under the perches or slats, in cracks or crevices, and particularly in the nest boxes.

Another form of mite burrows under the skin and roots of the feathers causing the disease known as depluming scabies, while a fourth form burrows under the scales of the leg causing the legs to become swollen and covered by a white, chalky deposit. This condition is called scaly leg.

A number of different species of lice occur in poultry attacking different parts of the body, i.e. the head louse, the body louse and the wing louse. The chicken flea, similar to that found on other animals, mainly attacks the lightly feathered parts of the back of the head and under the wings. Both fleas and lice can, if numerous, cause loss of condition and production by setting up irritation.

Several effective insecticides are now available in the form of dusts, dips or fumigants.

The dusts include sodium fluoride, pyrethrum, derris, D.D.T. and Gammexane*. The dusts are sprinkled into the base of the feathers and the treatment is usually repeated after fourteen days.

Gammexane, which is probably one of the most effective, is said to be non-toxic, but it is generally recommended that it should not be used on turkey poults under three months of age, or on sitting hens. Sodium fluoride can be used as a dip. Treatment by fumigation is usually carried out by painting the perches with 40 per cent nicotine sulphate, Gammexane or lindane.

More recently sprays from aerosols have become available, and are particularly effective with battery housed birds. Infested houses can be treated with creosote or paraffin emulsion, lime wash or paints incorporating D.D.T. or Gammexane.

NON-SPECIFIC DISEASES

There are a number of diseases associated with deficiencies of vitamins, certain minerals and other growth factors. The commonest are as follows:

VITAMIN A   A deficiency of this vitamin gives rise to the condition known as nutritional roup. The eyes show a watery discharge, which later becomes a white, cheesy deposit blocking the nasal passages and the eye sockets and producing small ulcers in the gullet.

Vitamin A rapidly loses its potency on exposure to air, and if fish oils are being used as the source of this vitamin they should be mixed fresh in the food. Most compound mashes today contain stabilised vitamin A. Green food, particularly grass meal, is another source of this vitamin. A deficiency of this vitamin also reduces hatchability.

VITAMIN B   There are a number of vitamins within the B complex and deficiency of several of them causes disease in chicks and particularly in poults. The condition known as curl toe paralysis, in which the toes turn inwards and in severe cases the chicks walk on the tops of their feet, is caused by a deficiency of riboflavin.

A deficiency of pantothenic acid and biotin, also members of the B complex, cause chick dermatitis, which is shown by small sores or crusty scabs on the corners of the beak and round the eyes, and on the feet of turkey poults.

---

*Some of these may not be allowed in certain countries and may not be available.

Vitamin A deficiency. Watery eyes, unsteady gait, ruffled feathers.

Rickets

Conditions Caused by Faulty Diets

The B complex, generally, is supplied by 3 to 5 per cent dried yeast, dried ;skimmed milk or dried whey, while the synthetic vitamins can also be used.

VITAMIN D   A deficiency of vitamin D gives rise to the condition known as **rickets**. The bones become soft and rubbery, the legs bend, the ribs thicken and affected chicks lose the use of their legs. The vitamin is supplied by 1 to 1/2 per cent cod liver oil, or by synthetic D3.

Rickets may also arise, however, as a result of an unbalanced or inadequate supply of calcium and phosphorus, while a lack of direct sunlight can also cause this disease.

CRAZY CHICK DISEASE

The exact cause of this condition is not fully understood, but it is thought that it may be associated with a lack of vitamin E in the breeders' ration, giving rise to chicks deficient in the vitamin, and the disease occurs when some other stress factors such as an excess of fish oils or possibly rancid materials are present in the mash.

It has recently been shown that an anti-oxidant called D.P.P.D. can prevent the condition. Infected chicks walk in circles, twist their head backwards over the body, fall over and paddle with their feet.

PEROSIS

This disease is also known as slipped tendon and is caused by a deficiency of manganese, probably associated with a deficiency of choline, which is part of the vitamin B complex. In the affected chicks the large tendons of the legs slip outwards from the joint, pulling the leg sideways, and the chick walks on its hocks. Usually the addition of 5 oz of manganese sulphate per ton of food corrects the disease.

Birds Affected by Sunlight Available
Top: Reared outside (normal)
Bottom: Absence of sunlight.

CROP

NORMAL HEN

CROP

ABNORMAL "CROPPY" HEN

Crop – Normal & One requiring Attention

6

FUNGAL DISEASES

ASPERGILLOSIS

This is also known as brooder pneumonia and results from chick inhaling the spores of the fungus Aspergillus, which is present in damp or mouldy hay or feeding stuffs. Chicks show symptoms of difficult breathing and on post-mortem small white abscesses are found on the lungs and in the air sacs. there is no known treatment of value apart from the removal of the cause and thorough disinfection.

MONILIASIS

This disease is mainly seen in young turkey poults and is caused by the fungus Candida, which causes small ulcerated patches in the crop. the condition appears to be associated with other factors such as malnutrition or faulty management, and losses can be high. There is, again, no effective treatment known and control depends on slaughter and disinfection.

CONSTITUTIONAL DISEASES

There are a number of constitutional disturbances which occur usually in individual birds, such as impacted crop, internal laying, impacted oviduct, prolapse, bumble foot, etc.

IMPACTED CROP

This may occur as a symptom of paralysis or pullet disease, or result from mechanical causes such as long, coarse straw, feathers, etc. It can sometimes be treated by holding the bird head downwards and massaging the crop. In advanced cases, surgical removal of the contents through the crop wall may be required.

In large flocks single cases are usually culled, but if several birds are affected the management, and the possibility of other disease being present, should be investigated.

EGG PERITONITIS
This is one of the commonest conditions met with in poultry in individual birds and often results from "internal laying", yolks passing directly into the abdominal cavity, or as a sequel to egg binding. Carriers of B.W.D. or fowl typhoid are often affected with peritonitis. Affected birds have a penguin-like appearance with a swollen abdomen. There is no cure.

EGG BINDING
This occurs when the oviduct becomes obstructed by a large, broken, or misshapen egg. Affected birds repeatedly visit the nest without laying and will be seen to strain. In some cases the impacted mass can be removed manually through the vent.

PROLAPSE
Prolapse of the vent may result from egg binding, inflammation of the oviduct or under the strain of heavy production. The condition is shown by the appearance of a mass of reddish tissue protruding from the vent.

Affected birds should be immediately removed since cannibalism frequently follows. Treatment can be attempted, but is not always successful. It consists in washing the exposed part and attempting to replace it by gentle pressure.

VENT GLEET
This is a contagious disease spread at mating or by contact with infected nest box litter. The area surrounding the vent becomes ulcerated and covered by a foul smelling, cheesy deposit. Infected birds should be destroyed and disinfection of the nest boxes carried out.

Treatment can be attempted, provided the birds can be isolated, by swabbing the affected parts with antiseptic lotions and antibiotic or sulphonamide dusts or powders.

BUMBLE FOOT
This is an abscess in the region of the food usually resulting from a small wound. The food becomes swollen and painful. Treat

ment consists in surgical opening of the abscess and removal of the pus. Antiseptic dressings are then applied.

VICES

CANNIBALISM & FEATHER PECKING
The commonest vices are cannibalism and feather pecking, and both are often associated with some other factor in management such as overcrowding or lack of hopper space, and in some instances are thought to be associated with a nutritional deficiency. Once started the vice is copied by other birds which are attracted by the bleeding surface.

These vices are commoner under intensive management, and can be controlled by de-beaking* or by the use of hen spectacles. Injured birds should be removed immediately as they attract others, and the injured area should be dressed with Stockholm tar. It has been suggested that an increase in the salt content of the diet is of some value.

POISONING
Poisoning usually occurs from accidental access to carelessly placed rat baits and the commonest are arsenic, phosphorus and zinc phosphide. It occurs rapidly and there is seldom time to apply antidotes. Generally, death is so acute that no symptoms are observed and diagnosis depends on chemical tests.

FOOTNOTE
There are new diseases which have occurred , such as Gumboro Disease , so the poultry farmer must keep up to date by reading *Poultry World* or other magazine. Health Regulations must also be observed at all times.

The drugs given in the text were those available at the time of writing , but new developments may have brought about new treatments, drugs and vaccines and therefore the advice of a poultry veterinary surgeon should be sought when a serious problem arises.

---

* Trimming beaks always appears a barbaric method which is against the principles of the fancier.

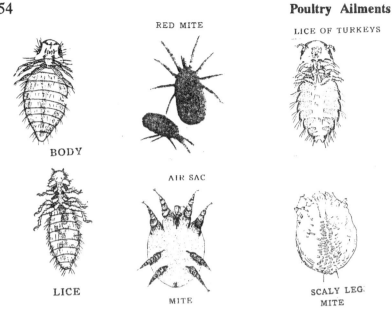

RED MITE

LICE OF TURKEYS

BODY

AIR SAC

LICE

MITE

SCALY LEG
MITE

Examples of Lice & Mite (after Kaupp)
magnified

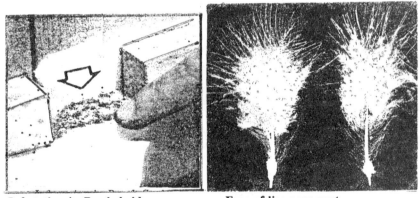

Infestation in Perch holder   Eggs of lice near vent

Lice & Mite

Dusting with insect powder or spraying, applying sulphur ointment or other
remedy is essential. Crevices and end of perches should be painted with creosote

PRACTICAL APPLICATIONS

DIFFICULTIES

The detection of a *specific* disease often presents great difficulties (but see end of chapter) and even a veterinary general practitioner has to carry out tests to determine the ailment. Often a post mortem examination is the only way. The poultry farmer or poultry fancier has even more problems because one complaint may be very similar to another.

The tabulation is a guide to what may be involved; remember that different diseases may show similar symptoms.

Part/Symptom	*Possible Ailment or Disease*
Abdomen swollen	Peritonitis, dropsy, white diarrhoea, damaged egg (see Vent).
Belching	Inflammation of **Crop**.
Breathing Erratic, gasping, noisy.	Gapes, pericarditis, air-sac mite
Choking	Poisoning, large piece of food in throat, irritant.
Comb pale	Air-sac mite, dropsy, infectious leukaemia, TB, white diarrhoea.
Comb dark; purple, mottled red	Blackhead, congestion of lungs, heart problems liver disease, pneumonia.
Comb yellow	Liver disease, visceral gout.
Comb pale then dark	Enteritis.
Comb with white scurf	'White Comb' ((Favus/fungal parasite)
Convulsions	Epilepsy, poisoning including accidental inhalation from pressurized can of mite killer.
Cough	Dusty conditions, bronchitis, diseases of respiratory system.
Crop enlarged & hard	Crop binding

Crop enlarged or soft	Inflammation, blocked with long grass, gastritis.
Diarrhoea	Multitude of possibilities; blackhead, fowl cholera, poisoning, unclean wa ter, roup, white diarrhoea, and any diseases of the alimentary (food digestion) canal.
Emaciations (going thin/feeble/wasting)	Aspergillosis, mites, TB, worms, visceral gout, white diarrhoea.
Eye sticky, discharging, face swollen, discoloured, pupil expanded.	Catarrh, roup, disease of the eye.
Face swollen	Roup, injury from fighting, frost bite.
Droppings green/discoloured.	Wrong food, lack of food, if bright green suspect cholera.
Fever, marked	Aspergillosis, infectious leukaemia, inflammation of oviduct, peritonitis.
Lameness	Aspergillosis, bumble foot, leg injury, rheumatism, scaly leg, TB.
Legs rough & scaly	Scaly leg mite.
Mouth discharging, gasping.	Congestion of the lungs, gapes, pneumonia.
Nausia & Vomiting	Inflammation of crop, poisoning.
Neck bent backwards	Wry neck, brain or nervous system damage, poisoning.
Neck limp	Limber-neck.
Paralysis	Apoplexy, heat prostration, internal growths (see Marek's Disease).
Saliva, copious	Poisoning.
Skin, puffed out in blisters	Emphysema, lice.

Skin scaly and scabbed	Body mange, favus.
Staggering	Leg weakness, paralysis, nervous system problems.
Thirst excess	Aspergillosis, peritonitis, liver disease, tape worm, excess salt in food.
Tongue hard & dry	Pip, respiratory organ disease.
Tumours/Lumps on head	Roup, chicken pox.
Wattles yellow (other indications also)	Fowl Cholera (also known as Pasteurel losis)
Vent red	Mite/lice infestation
Vent inflamed, tissue protruding	Prolapsus of oviduct.
Vent skin inflamed obnoxious smell	Vent Gleet

These symptoms may be spotted quite easily and treatment started. When deaths occur or what appears to be a serious breakout of an infection, the birds in the pen or flock should be separated from others and strict precautions should be taken to avoid the spreading of the disease.

What is important is for any sign of ailments to be spotted quickly and acted upon. Appearance is all important so any bird keeper should be able to detect a problem when it occurs. Signs are:

1. Birds does not move around.
2. May not take food or water.
3. May crouch or lie on its side.
4. Comb shrunken or discoloured (pale or dark).

5. Bird 'humped up'.

6. Feathers ruffled up or wings held low.

7. External parasites seen on body (near vent) or tiny holes in feathers or feathers eaten away.

8. Crop bulging and never emptying.

9. Bird visiting nest, but not able to lay (egg bound).

10. Droppings discoloured, indicating food problems or worms.

Many other signs coud be mentioned, but the reader should realize that a watch must be kept for anything unusual.

REGULAR HANDLING

As noted earlier, regular handling of birds is essential because this enables the handler to feel the breast bone, examine the face, look at the wings and remove any broken feathers. The bones near the vent, if open, indicate a hen is laying (two fingers between) and around the vent itself, which should be moist. Just above the vent the feathers should be clean and if faeces have stuck to the feathers these should be removed; if left eggs will be laid by the parasites and the bird will be infested with lice. If the faeces are soft and matted to the feathers this may indicate a disease and should be detected and treated.

ALTERNATIVE APPROACH

An approach relating to diseases follows on the next six pages. It was designed by a well known veterinary pathologist. Some cover the evidence from a post–mortem examination and may not be relevant for the average poultry keeper.

HOW TO RECOGNIZE DISEASE/AILMENTS IN POULTRY
BY REFERENCE TO BODY PART
Details of conditions may be found in various parts of the book.
Readers are advised to refer to the General Index on pages 95 and 96.

ADULT BIRDS (Still alive)

HEAD

Head twisting back, or on ground. Intestinal parasites; fowl paralysis;
lack of vitamins; acute indiges-
tion;

Limber neck; injury.

Rattling Noise in breathing. Catarrh of the windpipe; gape
worms; infectious tracheitis; roup.

COMB

White Patches or blisters. Usually Favus

Warts or Sores Fowl Pox
(also on eyelids and corners of
mouth)

Comb Dark Red, congested. Infectious tracheitis; liver
problems, especially if purple.

Comb Purple when normally red. Heart or liver problems; no
water.

Comb White or pale. Anaemeia (faulty nutrition?);
worm infestation; coccidiosis;
red mite; blood disease; TB.

Comb Yellowish Tumours of the liver; blood
disease called Eythromyelosis;
jaundice.

EYES

Yellow thick mucous; foul odour. Canker

Frothy discharge. Cold; canker; infectious catarrh.

EYES

Creamy accumulation in eyes.	Bacterial roup; cold; Avitaminosis A (lack of vitamin A).
Pupil small or irregular; blindness.	Fowl paralysis; worms; eye disease.
Iris grey or pale.	Lack of Vitamin A.
Bulging Eyes; eyes closed.	Glaucoma (dangerous disease – hereditary)

BEAK & MOUTH

Continually opening, neck outstretched.	Gapes; tracheitis;
Dribbling from mouth; birds dying.	Fowl plague; tracheitis.
Beak lumpy, edges irritated.	Thrush (an infectious rash).
Yellow matter on tongue.	Fowl canker. (Glycerin and antibiotics may cure)

NECK

Bare feathers; dry feathers; falling out. (Spraying with water containing a mild disinfectant is helpful)	Mite; lack of protein or greenstuff; feather pecking; excessive dryness (kept indoors).
Crop distended – gas/water/food.	Gizzard blockage; chronic coccidiosis; foreign body in crop.

BODY

Breast bone distorted and crooked.	Lack of Vitamin A; too early perching.
Going Light; thin; joints swollen.	TB; Tumours; coccidiosis; food deficiences.
Breast blisters or sores.	Unsuitable or lack of perching.

WINGS

Wings drooping or weak.	Cocccidiosis; worm infestation; injury; split wing (hereditary condition).
One wing dropped, weak.	Fowl Paralysis a possibility.

LEGS

One or both Legs paralysed.	Chronic coccidiosis; leg weakness (dietary); Marek's disease; fowl paralysis.
Lameness; weakness in leg.	TB; layer's cramp.
Rough & scaly legs.	Scaly leg (a mite); soreness esp. feathered legs.
Swollen joints; limping.	TB; lack of vitamin D.
Pads of feet swollen, limping.	Infection of foot from thorn, etc. or from landing badly (high perch). (Bumble Foot)

STERN

Abdomen swollen, drooping, featherless.	Egg bound; broken egg; burst oviduct; excessive fat; tumour.
Vent red and inflamed; bad smell.	Vent gleet; caecal worms; lice.
Rump featherless, bleeding.	Feather pecking; mite; diarrhoea.

Droppings streaked with blood; watery;   Coccidiosis;
evil smelling; greenish yellow.   poisoning; TB;
                                                 fowl typhoid;
                                                 cholera.

ADULTS (Post-Mortem Results after death)
Note: The diagrams given on pages 2, 11, 12, and 14, indicate the workings
of the systems and the main parts. Some fanciers carry out post-mortems,
but can be difficult and is better done by a qualified person.

GULLET & CROP

Crop filled with gas/water.   Stoppage in digestive tract.

Congested gullet & crop; boiled look.   Acute poisoning with corrosive
                                         poisons.

STOMACH & GIZZARD

White pimples on gullet   Avitiminosis; catarrh.

Gizzard filled with grass.   Lack of grit.

BOWELS

Thick cheesy dry nodules   TB in some form.

Dilated yet full of gas when killed.   Coccidiosis; poisoning;
                                        bad feeding.

CAECAL POUCH

Dilated, thin, transparent.   Chronic coccidiosis.

Cheesy masses.   Irritation due caecal pin-worms.

LIVER/SPLEEN

Enlarged, yellow nodules.   TB.

Dirty green 'bronze' colour.   Bacterial infection (salmonella
                                     pullorum?).

White spots, enlarged.   Lymphocythaemia or leu-
                                     cocythaemia.

White or red nodules (few)   Tumours

Liver Swollen, dark, enlarged.   Congestion due to disease.

KIDNEYS

Swollen, pale, nutmeg like.   Vitamin deficiences.

Shrunken, pale, rough surface.   Chronic lack of vitamins; nephritis.

Swollen, dark, bleeds freely when cut.   Congestion due to heart disease or acute infection.

OVARY

Not developed, yolks small & clotted.   *Salmonella pullorum.*

Ditto, dark red, yolks very small.   Avitaminosis D.

Swollen & pale, no yolks.   Tumours at times purple.

Eggs retained in oviduct, inflammation.   See under 'STERN' above.

LUNGS

Dark red, small pieces; do not float in water.   Broncho–pneumonia

Very pale or grey; on pressure oozes clear fluid.   Heart failure.

CHICKS (Chicks still alive)

High Mortality, white droppings.   Bacillary white diarrhoea

Open beaks, rattling noise, sneezing.   Gape worms.

Drooping wings, crouching, blood in droppings.   Coccidiosis, worms, parasites.

Huddled round heater, gasping.   Wrong heat/ventilation; brooder pneumonia.

Joints thick, weak in constitution, crooked breasts, lack of development.   Vitamin D and sunshine required.

Head twisted sideways.	Vitamin deficiency; indigestion; worms.
Leg bent outwards.	Slipped hock; when bird debilitated may be Marek's disease.
Feather pecking/picking.	Shortage of protein; overcrowding.
Toepecking.	Mite; overcrowding; spots on toes.

CHICKS (Post-Mortems after death)
Yolks not resorbed.

Incubation too high; disturbance during hatching; lack of vitamins in feed of parent; bacterial infection of mother.

Eggs not maturing due to bad yolks.

(yolks mixed with blood and other foreign bodies).

Many possibilities; faulty incubation.
(See *Artificial Incubation & Rearing* Joseph Batty, from BPH).

LUNGS
See above under Post-Mortems.

VENT
Pasted up or dry and pale.

Suspect BWD or other disease.

BOWELS
Congested, but empty, or
with white thickened spots.

BWD or Salmonells pullorum or fowl
typhoid.

LIVER & SPLEEN & KIDNEYS
See above under Post-Mortems. Usually
problems affect a number of organs.

OLDER CHICKS
Caecal pouches thick and dark.

Coccidiosis.

Vent red & inflamed.

Errors in feeding; lice; cannibalism.

WATCH FOR SYMPTOMS IN OLDER BIRDS
Sitting frequently in nests by birds.

Egg bound; yolk blockage;broody.

Shaking Head in jerky manner.

Chronic catarrh; mites in ear.

Squatting on ground constantly.

Burst yolk in abdomen.

ROUTINE MATTERS

TYPICAL OCCURRENCES

Keeping poultry is a delightful occupation whether as a business or as a hobby. However, problems do arise which are relatively minor provided the stock-keeper is able to deal with them. Many are of a routine nature and should be dealt with on a regular basis. Only birds which are fit and laying regularly should be kept and a target should be set to cover winter and summer figures. Those which fall short should be culled. The same applies to exhibition birds which may not be the best of layer, but should win prizes or be capable of breeding winners; any weak specimens falling short of standard requirements should also be culled.

Some of the matters which may occur are covered in this chapter.

HEAD & NECK

The head is a sure sign of the state of health. A bright comb, normally red, is an indication of condition. In a hen, it shows whether the bird is laying and in a male whether fit. The comb should be of the size which is natural for the breed (per Standards), but should never be over grown or 'beefy', because this is a sign of fatness or being kept in the wrong conditions - hens kept intensively often have overgrown combs of a pale colour.

A purple comb, except in a Sebright bantam or a Silkie, is a certain sign of being 'off colour'. It may denote a weak heart or some form of weakness. Many years ago I acquired an Indian Game cockerel from a well known breeder, who had managed to breed very heavy bone in to his birds. His legs were short and very stout, with a wide body, and had he developed he would have been a certain winner. Unfortunately, when he walked he gasped for and when he went to drink his face turned purple. The breeder had succeeded in producing a real heavy weight, but had gone too far and the bird was handicapped. Before maturity he simply dropped dead. The lesson is never try to breed exaggerated points to the stage where birds are unhealthy.

Heart, liver, or other organ malfunctioning may be indicated. Sometimes it is caused by neglect, from the birds not being given water, especially in hot weather. In the latter, be sure to supply shade and plenty of ventilation in sheds. Where there is an intensive system a failure in the air conditioning or the malfunctioning of fans can cause heat exhaustion and death. Poultry must have a plentiful supply of fresh air.

The comb can be treated with vaseline or other ointment, thus avoiding soreness or treating damage from scratches or from fighting. However, certain medicinal creams may cause 'skinning' and if a bird is to have cream put on the face and comb, prior to a show, make sure this is put on just before the event or the skinning process may have started and will spoil the exhibit.

When a comb is badly damaged or frost bitten it should be removed, a process called 'Dubbing'. This is also done for Old English Game and Modern Game males to give them a characteristic look, dating back to the days of cockfighting. The minor operation requires a pair of scissors, medium in size and curved, which are used to remove the comb and wattles in a specified manner. This is done at about 6 months of age and does not appear to give much discomfort; it heals very quickly, but for shows allow 4 weeks to heal.

The Eyes

The eyes should be prominent and clear. Any watering from the eyes, accompaned by blockage of the nostrils, will require washing with a safe liquid such as special eye drops or Optrex diluted with water. If a very thick substance develops in the nostrils and extends to the eyes (some form of Roup) the recommended bathing must be supplemented with anti-biotics and the affected birds isolated. This will need very drastic action and constant attention or the birds will have to be killed and incinerated. However, modern drugs can now cure this condition.

Cut off comb along dotted line

Trim wattles
and 'smooth'
off face.

Using curved
scissors cut
off parts
indicated.

The Process of Dubbing
From *Old English Game Bantams*, Joseph Batty

The Neck

Birds sometimes get lice or mite in the neck feathers. Dusting with insect powder may produce the desired result. Rubbing parts of the neck itself with sulphur ointment is a real deterrent. When the hackle becomes very dry and brittle it should be sprayed or washed, including a little washing up liquid and this can cure the problem which is always damaging to the chances of a Modern or OEG bantam when being shown.

The Beak

The upper mandible should fit nicely over the lower and the beak should not be too long. If a bird is kept too long in a pen for show training so the beak does not get worn down by use, it will be necessary ro carefully cut and file down to the normal size.

Upper Mandible should be trimmed when necessary: trimming knife & nail file

Ideal

Beak & Position for Trimming

THE BODY

The body should be of the **type** specified in the *standard*, complying with such terms as 'short in the back', upright in carriage' and so on.

The feathering will divide into 'hard feathering' and 'soft feathering' , referring to the tightness of the feathers around the body. In reality the distinction is not clear cut and, whilst hard feathered refers to Game breeds, there are many others which have tight feathers which are not strictly Game; eg, Sebright bantams. Even the soft feathered vary from being profusely feathered (eg, Orpington) to those with medium feathers (eg, Welsummers).

Besides complying with the *type* the specific bird must have as near perfect conformation as possible. Thus the following should be checked on a regular basis:

1. Breast Bone (Keel)

Should not contain indentations or bumps, which are a sign of too early perching or an unbalanced diet, especially lack of Vitamin D. At first signs take corrective action.

2. Back

Should follow a natural line across the body and not be indented or raised above the normal slight curve unless there is a tendency for this to occur in a specific breed; eg, Indian Game. Birds exhibiting such faults should not be bred from because the fault is a deformity and may be passed on.

3. Pelvic Bones (each side of vent)

These should be soft and pliable and sufficient space between the bones to allow an egg to pass through. Females not laying will have little or no space between the bones and at around 6 to 8 months the comb should redden and the pelvic bones open up. If this does not occur at the correct time the pullets in question may have to be culled and used as roasting fowl (provided there is no disease). However, with some breeds of exhibition poultry which produce very few eggs care should be exercised. Remember also that some bantams will only lay in the Spring and

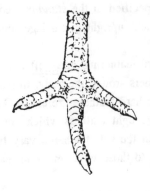

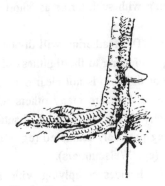

Twisted Toes   Duck Footed

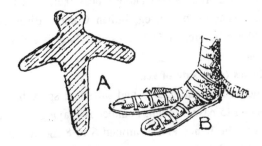

A Method of Straightening Toes
This is not always a success

The Legs

Summer months, although others lay quite well for much of the year.

5. Wings

The wings generally follow the contours of the body and the feathers should be strong and without defects. If on examination it is found that the main feathers are getting broken it may indicate that the diet is lacking protein. The hard feathered birds, such as Old English Game, not running outside with access to natural foods, and being fed only on corn are often brittle in the feathers, being easily broken This limiting of protein to keep the feathers hard can be carried too far and, in any event, will interfere with the fertility of the breeding pens. A compromise of layers' pellets and mixed corn (not wheat alone) is usually the answer.

If primary feathers are completely broken a bird cannot really be shown so the odd feather can be removed, allowing time for a new one to be grown for the shows; a few weeks will be essential. Any birds with split wings should be culled because these are no good for showing and will probably pass the defect to any offspring.

6. Tail

There is nothing worse than a poor tail and possibly of the wrong type*. Broken sickle feathers are the worst problem and take so long to grow. Birds which never seem to grow a tail are best culled. Watch carefully for lice and mite around the tail and vent. Remove any lice eggs from above the vent where they are laid in clusters. Rub around the vent with sulphur ointment and this will keep them away.

7. Legs

Legs should be of the appropriate length with straight toes and in Game birds it is important that the back toe is flat and straight back (not duck footed). Crooked toes are a nuisance and may be caused by diet or incubation. Often they straighten to some extent, but never very successfully and various methods have been tried to correct the fault. **Nails** should be short from constant use and if long, on birds running around,

* See *Poultry Characteristics, TAILS,* J Batty, where the different types of tail are discussed and explained.

this is a sure sign of an unproductive bird and may have to be culled. When birds are being kept in show cages the nails may grow long simply because there is no litter and earth for scratching. They should be cut to normal length, but not too short or or they will bleed. Nail scissors should be used.

If *scaly leg* is to be avoided, keep the legs clean and occasionally rub sulphur ointment into the scales. Cleanliness is important and breeds with feathered legs need special care; do not allow them to plod around in mud or in conditions which break the feathers and allow dirt to accumulate or sores will appear.

Some advocate using paraffin to cure scaly leg and related problems, but this is very harsh and may result in burns on the skin which are difficult to heal. In bad cases use a mixture of sulphur ointment and medicinal tar obtainable from a chemist; otherwise, sulphur ointment ny itself, well rubbed in, should effect a cure.

Nail Clipping When Birds In Training Cages

PROBLEMS WHICH ARISE

In keeping poultry of all types certain problems arise on a fairly regular basis; many are not ailments, but they do constitute losses and should be avoided or cured. They are now discussed.

EGG EATING

Hens and cockerels eat eggs which is a habit which should be stamped out. First an attempt must be made to avoid the problem starting. It may be because the birds have not been given water and turn to the eggs for liquid; or the shells are weak and break easily starting the habit – extra grit must be given. There may be a weakness in the diet – the author once loaned a bantam hen to a budgerigar fancier who allowed the hen to live in the bottom of the run scavenging from food discarded; it quickly started to eat its own eggs and could not be cured, yet when moved back with other bantams and fed normal layers' pellets the bad habit was forgotten and the following season laid eggs and hatched chicks.

Filling eggs with unpleasant contents such as mustard sometimes acts as a deterrent. An alternative is to have a nest box where the egg rolls away so the hen cannot poke at it. 'Pot' eggs can also be used in the nest. Ample nest boxing, kept clean with straw or hay provided, also help.

If an egg breaks in the oviduct this is a very serious condition and the hen usually dies. On the other hand, if the egg is intact, but the hen has difficulty in laying it will be necessary to try steaming the vent or lubricating by using a few drops of oil or vaseline. If the egg can be felt at the bottom of the egg canal it may be possible to put a finger into the anus and remove the egg most carefully, avoiding any damage to the channel. Should the egg break all the bits must be removed. A teaspoon of olive oil may then be given and the bird put on a soft diet of bread and milk so she can recover and probably lose some of the internal fat which is causing the problem.

Blood or meat spots in eggs can be a serious problem and downgrade any eggs affected. It may be due to malfunctioning of the oviduct as the

Feather Mite

This has been allowed to progress too far. Dipping in hot water with Jeyes Fluid added may be the answer, but do not make too strong or it will burn the skin. Never allow the head to immerse in the water or it may go in the eyes.

Ring Worm (Favus) in an Ancona Cock

Washing with disinfected water and treating with sulphur ointment will usually cure. Cleanliness is vital or the infection will spread.

egg passes down; plenty of exercise and a sound diet is a likely cure.

CANNIBALISM

This comes under various names such as feather plucking or pecking (or picking) and bullying. The habit is most frequent under intensive conditions when birds become bored and lack certain proteins or minerals in their diet. A dim light, and meat powder or fish meal in the diet, can help. Some poultry experts also suggest that mash (powder) fed in hoppers keeps the birds occupied and reduces boredom.

The author is not generally in favour of mash because it is very dry and unpalatable, although in intensive systems it may have its place.

The methods of de-beaking, a rather barbaric process, is also used as are 'specs' and rings on turkeys. Fortunately, with free range and semi-free range, practised by fanciers, the problem rarely arises. However, it is important to give fresh water and a balanced diet, with vitamins, protein and amino acids, supplemented by a plentiful supply of fresh greenstuff, and this will suffice to avoid the problem.

If birds have plenty of space any that are being bullied can escape so, again, natural poultry-keeping will give great benefits.

Allied problems are **toe pecking** and **vent pecking.**

Toe pecking usually occurs with incubator hatched chicks which appear to believe that yellow legs or a spot on the toe is food or is simply pecked by mistake and blood appears which causes great excitement and further pecking. Also when chicks or adults run in mud or earth which is sticky a ball of clay adheres to the toe, its removal may cause blood to come from a torn nail and, in turn, pecking starts. It is important to soak the ball before removal.

The wound should be treated with boracic ointment and then painted with creosote to deter further attacks.

Vent pecking may arise from the chick trying to deal with an irritation or from another chick seeing the vent and pecking it in the belief that it is food. Once blood is seen the bird affected will be attacked and many

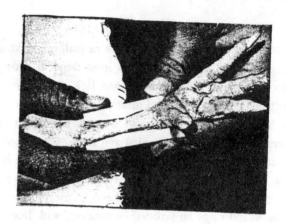

Getting Fracture in Position

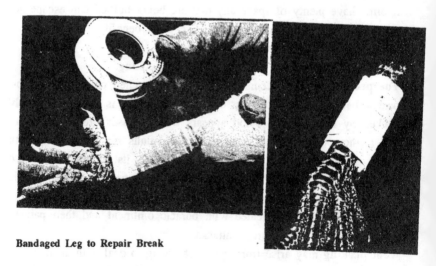

Bandaged Leg to Repair Break

Repairing Fractures to Leg

cases of young pheasants have been known where the poults have been disembowelled and almost eaten alive.

Any affected birds should be isolated and the wound washed with warm water and dettol. Dusting with boracic powder or a similar substance should be effected and only when cleared up should the birds be returned to the rest.

In all cases of Cannibalism the use of a dimmed light (if inside) and improved diet and a plentiful supply of Greens will be essential. Litter with mixed corn thrown in also gives the birds something to do. For preference the birds should be outside with plenty of space and fresh air.

CUTS AND FRACTURES

Any cuts sustained by the birds fighting or injury from other causes should be treated with urgency. The injured bird should be separated from the rest (a spare coop or shed and run should be kept as a 'hospital) and the wounds bathed with warm water containing an antiseptic. Any broken feathers are best removed. Then an ointment can be applied, such as Boracic, Sulphur or one of the newer remedies available from the chemists.

Broken legs can be repaired, but careful skill is needed to set the leg back in its original place; the normal positioning is essential. If the break is on the thigh the feathers should be removed.

With a simple, single fracture, the placing together of the bones is fairly easy. With a multiple fracture the task is more difficult and in very severe cases the bird may have to be killed.

Once in position, three narrow pieces of wood should be placed on a thin layer of cotton wool; these are the splints and foundation for keeping the leg in position. A cotton bandage should then be wrapped around the splints and these would be soaked in glue or Plaster of Paris, either of which should dry very quickly so that the leg is held rigid.

After 14 days the break should be healed and the bird will be able to resume normal life again; it should be kept by itself for a few days and then returned to normal.

Wing fractures can be treated in the same way and the wing strapped in position until healed.

MOULTING

Moulting is not a disease or ailment, but simply the process of a bird shedding its feathers and replacing them with new. Late Summer or early Autumn is the normal time for adult birds. It is important because the longer the moult takes the longer the period when hens do not lay. Moreover, some of the standard breeds seem to have difficulty with the process. Old English Game hens have been known to die half way through; Thai Game cocks, proud and majestic up to the moult, have simply dropped dead for no apparent reason.

The process can be accelerated or introduced earlier (important for shows) by putting the birds in warmer quarters and changing the diet to one which contains little protein. Once the feathers start to fall rapidly and new ones are sprouting the normal layers' pellets can be re-started. A supplement of cod liver oil or linseed oil may be added to the food and all this should add on the weight lost in the moult and bring them back into condiion.

The whole process should take 6 to 8 weeks and any birds which take much longer, unless show birds, might be better culled.

8

AILMENTS USUAL WITH THE FANCIER
DIFFERENT TYPES OF POULTRY

SMALL NUMBERS

Birds kept by fanciers are usually given personal attention, are kept in relatively small numbers, and are housed in small sheds with access to an outside run where 'free range' conditions can be present or simulated by the regular feeding of grass and greens and other essentials.

Many of the respiratory diseases and such problems as brittle bones come from being kept in close confinement with inadequate exercise and a diet which gives all requirements. The access to sunlight also makes a tremendous difference, eliminating many of the diseases by enabling birds to build up resistance.

Some diseases occur because birds *are outside* and it is these which must be watched. Thus scaly leg occurs when the birds are outside, particularly if there is fire ash in the run or other irritants.

Coccidiosis in young chicks becomes rampant when there is a damp spell and the chicks are outside; only by rearing inside (with sunshine coming in through wire netting - not glass) will the problem be avoided, and not even then if the disease is in the birds kept.

BANTAMS

Bantams are small poultry around 25 per cent of the size of large fowl of the same breed or similar. Some are true bantams and therefore have no large equivalent. Diseases suffered are much the same as those contracted by birds which have regular access to outside; they have built up anti-bodies so they are quite able to cope with any infection provided they are well fed and kept in hygienic conditions.

Pekin   Sebright

Old English Game   Rosecomb

Typical Bantams
Require management according to the type of breed.

Despite what may be imagined bantams are quite hardy and, once chicks reach about 10 days old they are lively and strong. However, certain breeds do suffer from problems. Those with feathered legs such as Pekins have to be watched very carefully or the feathers become dirty and the legs may become sore. Silkies suffer from the same trouble and are inclined to get scaly legs; some strains suffer from Marek's Disease (paralysis) so vaccination (Turkey Herpes Vaccine) is advisable. Coccidiosis in chicks, lice/mite, and Scaly legs are usually the most common ailments.

Sebright bantams are notorious for infertility and they may also suffer from Marek's Disease. Japanese bantams have very short legs and may have breeding problems; they belong to a type of fowl known as 'Creepers' and this condition can carry lethal genes if short legged birds are bred together.

Despite these problems, provided the houses/pens are given a good layer of soft wood shavings, which is changed regularly, many of the difficulties can be avoided. Bantams cannot stand cold, wet conditions so in dry quarters with adequate food and water the ailments are avoided. Some breeds such as Old English Game are quite hardy and require plenty of exercise, preferably with an outside run which is changed or dug over at regular intervals so the floor does not become matted with faeces and carry disease. Grass runs are the best, but should be allowed to rest and new runs used.

Modern Game, despite appearances are quite hardy except in very cold and frosty weather. Cases have been known of birds freezing to death, presumably because they are quite small and slight of build. The answer is to move them into a warm shed and ensure there is adequate food and water. In frosty weather, for all bantams, the water should be renewed – possibly with hot water to thaw out the founts – so that they have a drink each day.

Although not a disease the danger of rats must always be present. These dreadful creatures, in their quest for food, may suddenly appear and not only take the food, but also kill bantams and then leave the carcases after feeding on them. Concrete floors to the sleeping quarters are very desirable, as well as the blocking of any possible inlets to food, thus at least thwarting their efforts to get inside. Lay rat poisons under boxes or other places not accessible to the birds and the invading pests are eliminated. They must never be allowed to dig themselves into the foundation of a shed or the poultry are doomed.

DUCKS & GEESE

Fortunately ducks and geese are very hardy and suffer from few ailments. They have built up resistance to the common ailments. The main problems, discussed earlier, are as follows:

1. Hepatitis Virus

Young ducklings are affected. They sit with eyes closed and become very weak. A vaccine is available and may be used.

2. Worms

These may be of different types; the ducks or geese appear unhealthy and underweight, with little energy. Worm capsules can be used to eradicate the problem, but remember worm eggs may be picked up from grass land or other places where they roam so regular attention may be necessary. However, dosed too often is not good for the birds.

3. Goose Infuenza

There may be diarrhoea and the birds affected will breath with difficulty and gasp for breath. Anti-biotic drugs may help, but this is a dangerous disease.

4. Paratyphoid (Salmonella)

This has symptoms similar to Goose Influenza. Avoidance by using strict hygiene measures is the only way. Ducklings gasp for breath and suffer from trembling fits. They usually die fairly quickly.

5. Slipped Wing

The wing sticks out at the side, not wrapping around the body as usual. This is not usually a serious disease, but if suspected as a genetic weakness and not an injury, the birds should not be used for breeding.

6. Various Ailments

Soft-shelled eggs, lameness, difficulty in moulting and other minor ailments may also occur. Watering from the eyes can be avoided by supplying deep containers or a pond which allows the ducks or geese to immerse their heads. Acute cases are known as *White Eye* and antibiotics are essential. A very high protein diet should avoid *leg weakness* and *staggers;* for the latter, chopped onions are said to help.

White Eye

Examine Eyes & Head of Duck
for Signs of Problems

Slipped Wing – Does not Affect Health, but do not Breed if it
appears to be a fault rather than injury.

TURKEYS

Up to 10 weeks of age turkeys are very prone to problems, but after that
are quite hardy. Proper feeding with a balanced diet and taking care not
to allow the chicks to be exposed to the dangers is the best way of
combatting the diseases.
The major diseases in turkeys have been Blackhead and Coccidiosis,
but there are others.

1. Aspergillosis

Infection of the lungs. May be caused by cobwebs, dust, mouldy hay
and other irritants. Occurs in young birds up to 3 weeks although older
chicks can be affected. Prevention is only way to avoid by providing
correct conditions; ie, no dust, mouldy food, cobwebs, etc.
Signs: Gasping, weak, not eating.

2. Biotin Deficiency (Vitamin B complex deficiency)

Poor growth, loss of appetite, scabs around eyes, weakness of legs.
Usual 4 - 10 weeks. This disease can also affect the liver and kidneys.
A proper diet is the preventive measure.

3. Blackhead

Signs: Lack of appetite, sulphur-yellow diarrhoea, and darkened blue-
black face - hence its name, although not always present.
Usually at 4 - 12 weeks. Is a type of intestinal parasite which can live
for a long time in mud and dirt. Poultry usually carry this parasite, but
are not affected to the same extent, although they will pass the disease
on to turkeys. Very similar to Coccidiosis, but Blackhead has sulphur-
yellow droppings. Without treatment losses can be up to 90 per cent of
affected birds. The liver is affected by large greyish white areas.
Prevention: Drugs in food and also, once the disease occurs, give
prescribed drugs from Vet.
Cleanliness and clearance of any contaminated areas; rearing the birds
on slats or twilweld may avoid the disease.

4. Coccidiosis

Signs: Ruffled feathers, uneven growth of wing feathers., loss of appetite, not growing, become listless and thin, blood in droppings, die off
quickly.

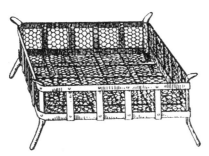

A simple design of portable destructor. Should the wire netting burn through, it ought not to be beyond the wit of man to replace it, for old fencing wire netting will serve the purpose.

Easily constructed Incinerator or Destructor.
These being readily erected, several of them can be placed in different parts of a large farm.

GAME BIRDS
(including Pheasants, Peafowl, Grouse, Partridges & Quail)

In the main the diseases of Game birds follow those of fowls, but with some variation in strength. For some there is not a great deal of experience available of life in domesticity. Pheasants can be extremely difficult to rear in large numbers; in small groups they are better. Cannibalism has to be watched very carefully, especially when the birds are just fully feathered; they may also panic with the result that many are suffocated.

An authority on pheasants lists the following diseases/problems:
NUTRITIONAL DISEASES
These include:

1. Rickets (lack of Vitamin D) which can occur up to 4 months of age – signs spreading of legs and thickening of bones along with depressed skull.

2. Curling Toes which is regarded as a Vitamin B deficiency.

3. Roup signified by discharge and then swelling and blockage of nostrils and eyes with a sticky yellow sudstance.

4. Six–day Chick Disease when chicks die off because of general absence of vitamins.

5. Perosis which is a disease of the leg bones – they become swollen at the joints and twisted. This is due to an excess of phosphates and is usually cured by a treatment with manganese.

PROTOZOAN DISEASES
1. Coccidiosis

Pheasants have their own strain. Chicks fluff up, eyes closed, and blood in droppings. Sulpha drugs and absolute cleanliness are the cure and future prevention.

2. Blackhead

Mopy and greenish yellow droppings (see Turkey diseases earlier). There is a cure, but carriers in poultry or turkeys are a real danger.

Perosis

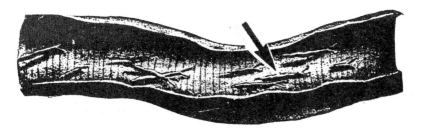

Gapes/Gape Worm

FUNGUS DISEASES

There are a number but the main one is **Aspergillosis** which is a growing mould which grows in the windpipe and lungs. Drugs such as Streptomycin can effect a cure provided the disease is caught early.

BACTERIAL DISEASES

1. Pullorum or Bacillary White Diarrhoea
The droppings are a liquid white and the birds are very dejected. All stock should be blood tested for the disease and carriers treated or killed.

2. Psittacosis or Parrot disease which affects many birds which then become carriers. Can be cured with antibiotics.

3. Fowl Typhoid and Paratyphoid
Deadly in pheasants so blood testing of stock is vital. Antibiotics can cure.

4. *Erysipelas* is a highly infectious disease causing drooping wings and tail and droppings in liquid form which are greenish yellow.

(e) Fowl Cholera where the head becomes badly swollen and there is difficulty with breathing, and the droppings are liquid. Usually birds have to be destroyed.

5. Avian Tuberculosis is a wasting disease. Pheasants are resistant to the disease, but it does occur. First signs are leg weakness and loss of weight. Carcases should be incinerated.

6. Navel Disease which is an infection of the navel; the bacteria, unless checked, go through to the body. Antibiotics in the food can help.

7. Botulism or Limberneck which is caused by decomposed food. Birds become weak in the neck (twisted) and legs and they stand dejected with drooping wings with periodic twisting neck and nervous activity. Modern drugs may clear the pouisoning provided caught in time.

VIRUS DISEASES

1. Air Sac Disease

This is fairly common and can result in deaths. The neck swells up and presses against the windpipe causing choking. Antibiotics in water can help the disease. Strict cleanliness and moving to new surroundings vital. Related to other respiratory diseases including Aspergillosis.

2. Newcastle Disease or Fowl Pest

A serious (notifiable) disease. The birds have running nostrils and eyes together with green liquid droppings; the neck is twisted and an affected birds gazes up to the sky. Vaccines can be used, but generally if the disease attacks all birds have to be destroyed.

3. Fowl Pox

Nodules on face and comb. Birds can be vaccinated.

PARASITIC DISEASES

1. Gape Worms

Pairs of small red worms that live in the trachea or windpipe where they block the passage way and cause choking (Gaping). If the birds manage to cough them up the worms or their eggs are swallowed by other birds and passed out in the excreta. In the soil they hatch out and are picked up by the birds directly or via affected earthworms or slugs, once more starting the cycle.

Various methods have been used to eliminate this deadly enemy. Fresh ground or ground which has been treated with copper sulphate will eradicate the disease. Treatments are also available, including a method of extracting the worms from the throat with a feather dipped in turpentine or the use of a special instrument – two people are essential for the process. Adding six grains of sulphate of iron to a gallon of water will kill any larvae which is coughed up. Various drugs are now available which bring relief.

Other worms are **round worm, tape-worm,** and **thread worm.** All can be eradicated by various drugs which are available.

2. Red Mite
A small grey insect which comes out of crevices at night and feeds on the blood of the birds. Creosote painted in the crevices is the best solution.
3. Scaly Leg Mite
Can be treated as described earlier.

4. Mange Mite & Feather Mite

5. Fleas
Insect powder will eliminate mite and fleas.

OTHER COMPLAINTS
All the other problems mentioned earlier for poultry occur in Game birds. Pulling the blood quills of feathers can be quite common and since pheasants very quickly become cannibals this problem must be watched and any signs of blood blocked out by using medicinal tar.

Specs for Avoiding Feather Pecking

POULTRY MEDICINES*

Poultry medicines fall into two main categories:

1. Vaccines for prevention;

(a) Live as a drinking water additive, by injection, or by aerosol spray ;

(b) Inactivated for injection into older birds, usually before egg laying starts.

2. Pharmaceuticals which are intended for giving treatment by adding to the drinking water or food. These are usually antibiotics or synthetic chemicals and are used to cure outbreaks of bacterial diseases or the elimination of parasites.

Only licensed persons or companies are allowed to develop and sell poultry medicines (must obtain a 'marketing authorization'). Products must be clearly labelled with detailed information and the dosage and frequency. There must be no danger of food or eggs being affected by the drugs. The date when the drug is out of the effective period should also be given.

The medicines are classified as follows:

1. Pharmacy and Merchants List (PML) which may be obtained from registered suppliers without prescription;

2. Prescription Only Medicines (POM) which may be obtained only on prescription. Generally these are food additives which are incorporated by the manufacturers. However, others are available only on prescription.

The fancier should stock only the medicines most likely to be needed such as Coccidiostat or ointments for various conditions. Remember medicines do run out of date and would be of little value if very old. Certain of the 'Old Fashioned' remedies do keep indefinitely and these should be kept im a medicine cabinet in an accessible place.

---

*Information kindly supplied by the National Office of Animal Health Ltd (NOAH)
3 Crossfield Chambers, Gladbeck Way, Enfield EN2 7HF

MEDICINES TO STOCK

Medicines or remedies should be kept in an airtight tin or in a cabinet which is kept closed so that dust or other elements do not get into the medicines.

If proprietary food is purchased in the form of chick crumbs, pellets or mash this will contain additives which are shown on a label attached to the bag. If purchasing in very small quantities ask the food merchant to supply a note on what is included. These additives give protection against certain ailments, such as Coccidiosis or Blackhead and only if there is a serious outbreak of a disease will it generally be necessary to give additional treatment. It follows that ailments such as Scaly Legs and the existence of parasites (lice, mite, fleas, and worms) must be watched very carefully.

BASIC MEDICAL SUPPLIES   (Kept in Medical Chest or Tin)
Scissors – 1 Normal; 1 Dubbing (medium curved, stainless steel)
Bandages
Sulphur Ointment (Large)
Dusting Powder for Mite, Fleas and Lice
Eye Drops
Vaseline
Disinfectant
Cotton Wool
Cotton Buds
General Germicidal Ointment
Plastic Bowl for bathing wounds, etc.

REMEDIES USED FOR POULTRY

There are many remedies which have been used by poultry men over generations; many are quite effective and are still available. Many are better treated with modern antibiotics although with viral diseases even they are not effective and it is a matter of waiting after taking appropriate action. Some conditions cannot be cured and birds should be killed and incinerated.

Basic Slag - For disinfecting ground; lime is also used for this purpose.

Bicarbonate of Soda - Relieves crop and gizzard problems; also Roup.

Boracic Powder - Toe pecking, Roup, Vent pecking, feather pecking.

Carbolic Soap - General cleaning.

Castor Oil - Egg binding, crop binding, diarrhoea, gape worms, limber neck.

Caustic Soda - Washing down shed walls, benches, incubators and room.

Charcoal - Settling digestion problems; gas in crop.

Cod Liver Oil - Source of Vitamins; leg weakness, lack of development. bronchitis, roup.

Common Salt - Roup, worms.

Creosote - Toe pecking, painting sheds and crevices (red mite). Avoid using on skin because it burns.

Epsom Salts - General pickup; assists with roup, layer's cramp, tapeworms, limber neck and worms.

Eucalyptus Oil - Roup and white eye.

Flowers of Sulphur - Dusting for lice and mite.

Formaldehyde - Disinfecting houses and incubators used according to instructions with no birds present. Usually a 2 per cent solution is used.

Glycerine - Favus and any scurfy conditions.

Iron Sulphate - Anaemia, roup.

Lard - Favus, scaly leg (sulphur ointment better).

Lime - Disinfecting ground, liming sheds (not much used these days for sheds).

Linseed Oil - Bumble foot, rubbing legs to avoid scaly leg.

Menthol Crystals - Wet roup and white eye (dissolved in hot water and birds breathe in vapour).

Olive Oil - Gape worms, prolapse, tape worms, wet roup, white eye.

Paraffin - Red mite (paint in crevices where they live); bad cases of scaly leg (but must never be put on skin - burns dreadfully).

Permanganate of Potash Crystals – Fumigation of incubator (see Formaldehide), bumble foot, wet roup, diarrhoea. (put in warm water until red).

Rhubarb, Tincture of. – Jaundice (yellow comb). 10 drops in a teaspoon of warm water.

Soap – Depluming scabies (wash with disinfectant in water).

Sulphur Ointment – Best all round remedy for mite, lice. wounds, scaly leg, favus, etc.

Tar, medicinal. – Scaly leg and bad infestations of mite, etc. May be mixed with sulphur ointment.

Thyme, oil of. – White eye, roup.

Turpentine – Gape worms, tape worms (mix with olive oil in equal parts and give 3 teaspoonfuls ensuring the bird takes the mixture which is unpleasant). Purge with Epsom salts before dosing with turpentine and olive oil.

Vaseline – General purpose for wounds, scurf, etc.

Vinegar – Scaly leg, roup.

Zinc Ointment – Dry roup.

Zinc Powder – Vent and feather pecking, wounds, dry roup.

After birds are treated it is important to treat the ground or move them to fresh ground. Where worms have been treated use carbolic disinfectant, such as Jeyes Fluid, to wash down sheds and kill any parasites or their eggs.

Dosages: Some are given above, but much depends on the ailment. Remember poultry weigh from about 2 kilos to 5 kilos so on body weight only a few drops of a remedy will suffice. Ointments are rubbed in and the amount required can be seen quite easily.

INDEX